Reassembling Digital Placemaking

This book presents digital placemaking as a new testing ground for urban democracy. It explores the participatory practices of digital placemaking and their implications on blurring formal and informal boundaries of decision-making and urban politics.

Drawing on examples from Australia, China and Taiwan, the book examines how contemporary cities are witnessing an era when digitality becomes a mode of placemaking and participation becomes an urban condition. Such developments manifest as urban activism, creative branding, tech placemaking and digital governance that trigger changes in urban engagement and politics. This book views digital placemaking through an assemblage lens to demonstrate how it can be a relational site of contentions and collaborations among civil society, industries and governments without observance of strict boundaries. Contrary to an often binary and zero-sum reading of urban politics, this book advocates for a tripartite assemblage model of urban politics that is neither hierarchical nor deterministic. This book proposes an ouroboros model as a systematic approach that conceptually anchors digital placemaking studies to the nexus of urban institutions and digital technologies. It reconceptualises urban politics as a relational process of nuances, contingency and complexity in the flux of cosmopolitan power movements and inquiries. Ultimately, the book develops the notion of socio-technical natality to counter the myth of tech inevitability and instil a thesis of hope and change.

This book will interest researchers, policymakers and professionals in urban studies, media studies, cultural studies, urban communication, creative placemaking, community engagement, urban advocacy and urban governance.

Isabel Fangyi Lu is a Senior Project Adviser at the Department of Education, Victoria, Australia. She is an Honorary Fellow of the School of Culture and Communication at the University of Melbourne. Having obtained her PhD from the University of Melbourne, she pursues interdisciplinary research and public projects at the intersection of technology, arts and built-environments.

Routledge Research in Place, Space and Politics
In memory of founding series editor Professor Clive Barnett,
University of Exeter, UK

This series offers a forum for original and innovative research that explores the changing geographies of political life. The series engages with a series of key debates about innovative political forms and addresses key concepts of political analysis such as scale, territory and public space. It brings into focus emerging interdisciplinary conversations about the spaces through which power is exercised, legitimized and contested. Titles within the series range from empirical investigations to theoretical engagements and authors comprise of scholars working in overlapping fields including political geography, political theory, development studies, political sociology, international relations and urban politics.

Spaces of Tolerance
Changing Geographies and Philosophies of Religion in Today's Europe
Edited by Luiza Bialasiewicz and Valentina Gentile

Geographies of the 2020 US Presidential Election
Edited by Barney Warf and John Heppen

Power and Space
Essays on a Shifting Relationship
John Allen

Uncommoning, Difference and Politics
Wordless Production of Paraguay
Antonio August Rossotto Ioris

Our Utopian Futures
Imagining a Sustainable Dialogical and Inclusive World
Valentin Mihaylov

Reassembling Digital Placemaking
Participation and Politics
Isabel Fangyi Lu

For more information about this series, please visit www.routledge.com/series/PSP

Reassembling Digital Placemaking

Participation and Politics

Isabel Fangyi Lu

Routledge
Taylor & Francis Group
LONDON AND NEW YORK

First published 2025
by Routledge
4 Park Square, Milton Park, Abingdon, Oxon OX14 4RN

and by Routledge
605 Third Avenue, New York, NY 10158

Routledge is an imprint of the Taylor & Francis Group, an informa business

British Library Cataloguing-in-Publication Data
A catalogue record for this book is available from the British Library

ISBN: 978-1-032-52704-8 (hbk)
ISBN: 978-1-032-52706-2 (pbk)
ISBN: 978-1-003-40797-3 (ebk)

DOI: 10.4324/9781003407973

Typeset in Sabon
by SPi Technologies India Pvt Ltd (Straive)

Contents

	Acknowledgements	*vi*
1	Introduction	1
2	Theorising Digital Placemaking	25
3	Activist Assemblage	49
4	Organisational Assemblage	70
5	Governmental Assemblage	92
6	Conclusion: Reassembling Digital Placemaking	115
	Index	*131*

Acknowledgements

Aligning with assemblage thinking, I acknowledge that all encounters and experiences have shaped my worldview and informed this book. I am indebted to many intellectuals and institutions throughout my PhD candidature, academia and government.

My PhD supervisors, Scott McQuire and Nikos Papastergiadis, exemplified erudition, rigour and life-long commitments to intellectual pursuits. Their guidance over the years continues to be invaluable. Clarifying ideas and interrogating writings are liberating events for which I am deeply grateful. My dissertation examiners, Germaine Halegoua and Rowan Wilken, provided enriching insights.

The book owes much to the knowledge and generosity of my interviewees and interlocutors, who enriched the Australian narrative and expanded the book to Chinese and Taiwanese contexts. These include contacts across academia, advocacy groups, activist organisations, tech start-up and art communities, place management, consultancies, engagement sectors, local and state governments. Special thanks to Deborah Allanson, Oliver Armstrong, Donald Bates, William Blakeney, Kate Brennan, Danny Butt, Gemma Campbell, Paul Carter, Trent Clews-de Castella, William Coogan, Lennon Yao-chung Chang, Helen Christensen, Joseph Cox, Jonathon Daley, Tania Davidge, Mark Davis, Rui Ding, Ceren Ergenc, Jennifer Frean, David Goodman, Glenn Harding, Robert Hassan, David Henry, Matthew Jones, Crystal Legacy, Dale Leorke, Rohan Leppert, James Lesh, En-Chi Lin, Cao Liu, Paul Long, Tony Luo, Natalie McGlone, Tongzhu Martha Meng, Tony Moore, Niamh Moynihan, Matt Novacescki, Elizabeth Parke, Dragan Pavlićević, Jasmin Pfefferkorn, Bruce Ramus, Anubha Sakar, Ian McShane, Jens Sinn, Bree Trevena, Jian Xiao, Danielle Wyatt, Wilfred Wang Yang and Audrey Yue.

The Australian Research Council's Discovery Project, *Arts Precinct, Cultural Participation and the Networked Public Space* (DP170102796), has supported this research. The School of Culture and Communication at the University of Melbourne has supported me as an honorary fellow. I thank the Networked Cities Research Network, Urban Assemblage Conference, iNetworkSociety and PhiloMedium for providing platforms for urban communications research.

The book also benefited from the professionalism and generosity of Routledge editors Faye Leerink, Akansha Bharali and Stephen Poole.

Thanks to my Victorian departmental colleagues, particularly Shura Shukor and Brendan Rigby, for fostering environments conducive to creativity and reflection.

Gratitude to my friends for their inspiration, strength and comraderies: Nima Behroozi, Bo Chen, Joyce Cheng, Xian Feng, Wendi Li, Ella Relf, Chaohui Xie, Lili Wang and Karin Zhu.

To Andreas, who reads everything I produce and tempers my writing ambitions with German romanticism and alarming images of the *Eierlegende Wollmilchsau* (egg-laying wool-and-milk-giving pig, an unachievable ideal of perfection) and *Phrasendreschmaschine* (words generator that spits out smart-sounding rubbish).

For the lateral thinkers and actors of the cities.

1 Introduction

In recent decades, cities globally have experienced rapid digital transformation. Navigating spaces, conducting transactions and connecting with others have become almost unthinkable without internet-enabled smart gadgets. While public opinion on the impacts of digital technologies and social media platforms varies, discussions around urban technologies often become 'infrastructurised'. China, where I grew up, is a pioneer in establishing national standards for smart cities and has begun testing a smart sustainable community in Beijing. At the start of 2021, a debate in a WeChat discussion group caught my attention. The trade-off between citizen rights and technological benefits is at the heart of the issue.

A digital transformation researcher argued that privacy concerns are clichéd. In his understanding, further discussions on abstract concepts were unnecessary. No one could try to halt the real existing smart cities. Therefore, the focus of current discussions should be on maximising technological benefits and minimising risks. An academic refuted this, questioning whether the public had willingly traded citizen rights and freedoms for convenience and safety or lacked a voice or participatory channels. Citing the recently abandoned deal between Alphabet's Sidewalk Labs and the City of Portland, the scholar argued that 'technology adoption should not solely be a governmental choice but a negotiation among citizens, tech companies, and local governments.'

The academic was then told that if he wanted to have a voice, he should 'participate in the National People's Congress'.[1] This effectively ended a constructive conversation. The imagined affordances of digital technologies theoretically bring more participatory politics, but in a country that lists 'the coverage rate of smart surveillance'[2] as a governmental assessment standard and unabashedly uses technologies for mass control, there are practical obstacles for a critical-minded scholar to 'participate' in China's highest organ of state power.

Any media scholar critical of technological determinism could contend that digital cities are far more complex than neutral instruments to be tuned on demand. The researcher demonstrates a technophilic determinism that sees no need to question the provenance and rationales of new technologies

DOI: 10.4324/9781003407973-1

as they are already an urban reality. The 'taken-for-grantedness' of urban technologies—their reduction to static and mundane infrastructures—allows them to escape public political scrutiny. More parents track their children's 'screen time' than equip them with anti-facial recognition masks. This attitude is prevalent among local governments that eagerly adopt the urban digital transformation, following Silicon Valley trendsetters and marvelling at the next shiny tech object. Reducing technology to its technical and operational aspects is a technocratic tactic portraying smart cities as an *a priori* mode of urban administration that can be operationally fine-tuned but never fundamentally challenged. Looking back, political scientists and observers will regret if they ever have downplayed the political ambitions of technologists and entrepreneurs and neglected the signs and manoeuvres of power grab that seeps through from immaterial to material spaces.[3]

It bemused me how the researcher stifled a valid political critique by normalising existing political orders and depoliticising urban technologies with such political righteousness. This uncritical mentality on technology is more harmful than systematically censoring political criticism. However, I aim to avoid the traps of technophobia that ignores historical and technological progress and technophilia that disregards the instrumentalisation of digital technologies by capital and the state. This book explores an open way to unpack politics rather than rushing into political judgements, asking: *How can we inquire into the politics of digital cities?* The answer lies in comprehending the strategic lens of digital placemaking.

1.1 Inquiring into the Politics of Digital Placemaking

Urban spaces have been historically mediated, and cities have functioned as media. Rife with public and communicative infrastructures, contemporary cities cast impacts beyond their physical layouts through digital assets, data logic and mediated presence. Urban life and its systems of checks and balances undergo constant institutional and cultural transformation. Lewis Mumford (1938) suggested that the shift from village agriculture to urban institutions further capacitated a scaled-up area of local intercourse. Jürgen Habermas's (1991) influential account of the bourgeois public sphere emerged from rational-critical debates in semi-public cafés and salons. Benjamin Bratton's (2016) further claimed that today's computing systems form a multi-layered, global megastructure that is replacing other forms of governance and sovereignty with political consequences manifesting on a planetary level. Media technologies and systems have a profound and enduring impact on urban socio-political ordering and freedom.

This book begins by acknowledging the ubiquity of digital media technologies in contemporary cities and their profound political implications. We employ digital technologies as tactical tools for designing, representing and modifying our surroundings: computer-aided design, virtual reality for showcasing real estate and tourism, and livestreaming platforms enabling targeted urban branding through data analytics. Building façades and interiors are adorned with media screens to create attractive and interactive destinations amplified through social media platforms and tourism influencers. Often unknowingly, we traverse environments replete with sensors, surveillance cameras and artificial intelligence (AI)-equipped robots in newly fashioned smart precincts, leaving digital traces and data exhausts to be mined, modelled and analysed for creative, commercial or policy products.

Digital technologies also inspire new decision-making and socio-political structuring, enabling urban forms previously deemed impossible. In Hangzhou, transportation systems and public affairs increasingly rely on Alibaba's AI-powered City Brain, which analyses real-time big data from numerous urban sensors and surveillance cameras. The Saudi government is tendering an ambitious master plan for a linear smart city boasting no cars, streets or carbon emissions. Mobile and locative technologies facilitate self-organisations, increasingly distributed through blockchain and decentralised autonomous organisation (DAO). Public life has experienced new mobile and visual paradigms and witnessed new forms of sociality, conflicts and collective actions. Technologically mediated cities bring about disorienting changes in human affairs and political realities.

One strategic way to apprehend this changing landscape is through the emergent notion of 'digital placemaking'.[4] Placemaking speaks of how people shape spaces to create meaningful experiences (Hes et al., 2020) and predicates the notions of community and participation. Conventionally viewed through the lens of place creation, this book extends placemaking to include moments of maintenance, inhabitation and governance. In this book, I will demonstrate how placemaking integrates various industries of the contemporary urban environment, such as planning, designing, management, governance and inhabitation. As a trans-disciplinary concept, 'placemaking' appeared in urban studies in the 1960s and was later adopted as 'creative placemaking' by the creative industries and as 'digital placemaking' in media studies. Although the human history of making places predates the emergence of cities, placemaking as a professional practice originated in the 1950s and 1960s amid massive urban transformations. Urban planners and designers sought to rediscover public spaces and re-establish a 'sense of place' to foster community experience and human needs for belonging, safety and orientation (Schneekloth & Shibley, 1995).

From the 1990s onwards, state and local governments globally took on the role of placemakers by developing creative policies that leverage arts, culture and innovation to activate place attractiveness and stimulate tourism. By the 2000s, digital placemaking emerged as a multifaceted concept encompassing public-facing business strategies, media projections, urban interfaces and informatics, as understood by marketers, media architecture practitioners, human–computer interaction (HCI) designers and scholars. Digital placemaking provides a holistic lens for understanding digital cities, encapsulating key urban domains like architecture, culture, economy, politics and technologies.

The notion of digital placemaking crystalises what the Club of Rome defined as the problematique (Meadows et al., 1972, p. 13), a meta-system of problems with the potential to counter the 'fragmentation of reality'. While digital placemaking and smart city initiatives leverage technology to improve urban living (quality, efficiency and sustainability), their foci diverge. Smart cities tackle broad infrastructure and service systems, while digital placemaking prioritises fostering connections and emotional vibrations between people and places, often at a more scalable and adaptable level, evoking deeper explorations of the human condition and relationship to technologies.

This book focuses on the political implications of digital placemaking, through which I uncover new modes of public participation and urban politics. Henri Lefebvre (1970/2003) advocated using urban society as a base of theoretical thinking and political actions. Cities manifest as macro accumulation of micro actualities and are privileged sites for studying the 'commoning' of our worlds. The politics of digital placemaking concerns the composition of digital cities, political realities and the processes of urban interactions. First, our actions of building cities and making places are inextricably embedded within the intricate networks of 21st-century neoliberalism and geopolitical competitions that fetishise efficiency and profitability. Understanding the politics of digital placemaking offers a reflective pause, allowing us to contemplate beyond immediate milestones like launching a new service or inaugurating a new smart precinct. Additionally, traditional compositions and political attitudes are being disrupted in this turbulent decade with the rise of deglobalisation, authoritarianism, military adventurism, climate crisis, global pandemic, sluggish economies, populism and 'post-truth' politics. Understanding the politics of digital placemaking helps us re-establish an enlarged sense of community and reorient ourselves towards a common goal of co-existence.

In this book, I am careful in applying two existing theoretical assumptions about the politics of digital cities. These assumptions, technological determinism and a Eurocentric critical urbanism, create a binary that hinders a nuanced understanding of the relationship between the urbanity, technology and digitality. Technological determinism, either techno defeatism or techno hubris, portrays digital technologies as an inevitable force shaping urban life, with a preordained political outcome. I avoid partially rehearsing accounts that emphasise the effect of media technologies on digital placemaking as a liberalising force enabling decentralised access, individual choices and urban collaboration (e.g., Foth et al., 2015; Fredericks et al., 2016; Hespanhol et al., 2017; Latorre, 2011; Offenhuber & Ratti, 2014; Polson, 2015; Tomitsch, 2016), or a deteriorating power leading to surveillance, polarisation and seclusion (e.g., Barns, 2020; Graham, 2020; Kitchin, 2014, 2015; Leszczynski, 2020; Lynch, 2020; Sloterdijk, 2016; Varoufakis, 2024). These narratives may be valid on their own account, but a greater need exists to avoid essentialising and decontextualising technologies (Kellner, 2021) and develop a theory that evaluates their impact with granularity, by case and by scenario, to inform future actions.

I further avoid emulating the established research path of critical urbanism on two grounds. First, the development of critical urbanism in a Marxist tradition is intertwined with the Eurocentric historical process of modernisation (de Munck, 2017), which overlooks the diverse experience and contexts of cities around the world and risks reproducing narratives of modernity. In addition, critical urbanism, often focusing solely on critiquing the capital accumulation in urbanisation, can reduce cities to political-economic entities that predominantly convey the agency of money, rationalisation and neoliberalisation (Sadowski, 2021). It assumes the technological as an external force being exerted upon a pre-existing political battle on the urban stage, with the governments and megacorporations on the one side and citizens on the other. It, therefore, creates an 'either-or' scenario that falsely forces urban planners and policymakers into a dilemma: embrace technological progress and economic growth at the expense of civic interest or prioritise community needs and risk stagnation.

This book proposes a new approach to inquiring into the politics of digital placemaking by tracing the relationships that comprise the political. Drawing on concepts from science and technology studies (STS) and assemblage theory, I argue that the political is not a singular force but rather an emergent property of the complex interplay between various actors, technologies and socio-cultural codes. For Colin McFarlane (2011), a dialectical approach aims to subvert the social totality by uncovering power relations in opposing forces, while an assemblage acknowledges the symbiosis of different power relations and seeks to discover possibilities of creating new 'lines of flight'. The assemblage approach focuses on tracing the trajectories of controversies surrounding digital placemaking practices. By examining the emergence, evolution and resolution of these controversies, we can identify unexpected alliances and unveil new possibilities for urban action, knowledge and democracy.

The inception of a new politics of digital placemaking involves a set of real political shifts, realignment of state, institutional and civic power, and the rise of participatory and ubiquitous media. Since digital placemaking is an emerging and practice-led field, theoretical research must overcome multiple obstacles. These include defining digital placemaking from myriad empirical applications, disentangling it from the smart city narrative and finding an intellectual home that honours its interdisciplinarity. In the following, I interweave four emerging theories of city, media and politics into my narration of digital placemaking and propose a way forward. After discussing existing models of urban public participation, I propose a theoretical model to facilitate my inquiry into the politics of digital placemaking as a rebuttal of political nihilism and technological defeatism.

1.2 Four Horizons for Digital Placemaking Studies

Four theoretical trends enrich the notion of digital placemaking and unfold new horizons for its research. First, digital placemaking does not merely create a tokenised reflection of the city but is integrated with it. Scott McQuire (2017) observes that digital media technologies have become a pivotal modality of contemporary placemaking, with ubiquitous, location-aware, convergent and real-time 'geomedia' embedded in the urban façades, programs and operations. We are now accustomed to transitioning between and integrating into different media and temporal paradigms as they co-exist and entangle (Mattern, 2017). During the Covid-19 pandemic, many of us became attuned to the regime of digital health, faithfully adhering to conditions of entry by scanning every visible QR code. This behaviour arises not because a two-dimensional code deters our three-dimensional being but from an almost blind trust in the system that somewhere someone will ethically merge our locative data with health data, inform those potentially exposed to the viruses and enhance community safety.

In essence, digital placemaking transcends a mere array of new practices and embodies deeper layers and flows of actions, logic and consequences within digital cities. There is danger in reducing digital placemaking to a specific code, application, device, function or program. It should be comprehended as an 'ensemble' within the complex urban interactions that traverse science and society alongside belief, superstitions, bureaucracy, norms and morality. We begin to observe normalised paradoxes: for instance, why so many citizens condemned Alphabet's (Google's parent company) smart city plan in Toronto as mass surveillance and neoliberalism while comfortably using Google Maps or holding Google stocks. These digital ambiguities paralyse our intrinsic judgement of whether a digital placemaking instance is empowering or alienating, good or evil. Instead, we must unpack myriad phenomena alongside 'the institutional settings, legal and regulatory environments, business models and cultures of use that instantiate it' (McQuire, 2017, p. 161). The renewed interest lies in discovering more instances of digital placemaking and following their interactions over extended periods to comprehend their real-world implications.

Second, assemblage urbanism offers a new theoretical foundation for examining digital placemaking as a practice of actors, relations and processes, enacted through heterogeneous urban elements (e.g., objects, entities, affect, capital, practices, norms, values and symbols). Places are scalable and socio-material ensembles; as Yi-Fu Tuan (1975, p. 165) famously proclaimed, '[p]lace, at all scales from the armchair to the nation, is a construct of experience; it is sustained not only by timber, concrete, and highways, but also by the quality of human awareness'. Tim Cresswell (1992) further dissected a 'place' down to three compositions: location (place whereabouts), locale (material setting of social relations) and sense of place (feelings and emotions a place evokes). Invoking Deleuze and Guattari's *engencement* (1987), Ignacio Farías and Thomas Bender (2010) conceptualise the urban as a de-centralised 'multiple object',[5] 'a multiplicity of process of becoming, affixing socio-technical networks, hybrid collectives and alternative topologies' (p. 2). Different components, such as urban life, traffic, goods, money, plans, policies, nature and infrastructure, amalgamate to give intensity and tension to the streets. Different assemblages converge, concretise and activate different senses of place.

Therefore, digital placemaking is not neatly arranged on a tabula rasa but is animated by the polysemic and multifaceted urban fabric. Simultaneously material and cultural, digital placemaking coalesces affect, discourses, technologies, practices, and institutions. The fireworks and digital display from Taipei's 101 building on New Year's Eve simultaneously serve a tourist attraction, a profitable business case for the Taipei local government, a mass traffic jam and environmental hazard for some locals, a surveillance zone for security guards on stand-by, a private memory for a young man who has just proposed, and a geopolitical theatre showcasing Taiwan's soft power. Digital placemaking can be understood through a rhizome that ceaselessly establishes connections between heterogeneous components through semiotic chains, power organisations, art, science and social circumstances. It is non-hierarchical and multi-scaler and can unfold into multiple artefacts, flows, institutions and systems that undergo constant movement. An assemblage view liberates digital placemaking from being studied as discrete aspects (e.g., media façade, urban event and user interface) or project phases (e.g., urban planning, designing and policymaking) but as a course of actions to follow and consequences to examine.

Third, the prevalence of participation enables digital placemaking to be examined as a mundane activity of contemporary digital cities. Barney et al. (2016) argued that participation (i.e., taking part in something with others) has become environmental and normative. New media are necessary but insufficient for the broader participatory condition, spanning polity, business and culture. This extended notion of participation becomes routinised in our everyday social, economic, cultural and political activities, from the Occupying Movement to mass surveillance, from open government initiatives to interactive art installations. Though public participation is unevenly distributed and variously implemented, digital placemaking increasingly showcases urban politics performed outside the electoral circle. In other words, participation is no longer occasional and ritualistic but is mundane and ambient in its potential to be realised. Different forms of government (at least nominally) acknowledge various forms of public participation as a binding principle of right actions. Even regimes that lay no claim to parliamentary democracy have their own models or narratives of citizen participation (e.g., China's 2021 Holistic Living Community Building Guide).

On a global scale, participation as a prioritised state of affairs has benefitted from lowered technological affordances and a change in public governance philosophies. Concurrent with the wide adoption of Web 2.0 in the late 1990s, New Public Governance (NPG) rose as a public service delivery approach that rejected seeing the governments as the sole decision-maker but drew on the skills and resources of public, private and civil society sectors through distributed and autonomous networks (Osborne, 2005). While the participatory condition implies a potentiality, it does not ensure that opportunity will automatically instantiate nor offer quality assurance. The ease and appetite for community participation have seen the dazzling proliferation of new participatory digital placemaking practices, such as design hackathons, urban activism, participatory branding and community engagement. As these participatory practices stabilise and formalise, we rarely question the validity of the participatory principle, focusing instead on its implementation. In other words, the old scepticism around the 'what's and 'why's [is] gradually substituted by the 'how's: how to improve the inclusiveness, quality and efficacy of participation and how to counterbalance it with the Tocquevillian 'tyranny of the majority'. As urban politics extend beyond the vicinity and immediacy, new constituencies, communicative forms and binding principles ensue. The participatory condition is the sword of Damocles, akin to Bernard Stiegler's (2010) notion of pharmakon,[6] simultaneously poison and remedy, power and peril. Digital placemaking is a unique site to comprehend the new hopes and anxieties of this deafening era where everyone, in principle, can have a voice.

Lastly, the emerging articulation of urban cosmopolitics enables a political understanding of digital placemaking beyond localism and West-centricity. Digital placemaking can be local displays of planetary issues, as distance ceases to be a significant obstacle for collective actions and civil movements. In recent years, a global counter-current has been witnessed against big tech's transformation of urban public space. Examples include public protests against Amazon's headquarters in the City of Long Island (USA) and the Apple flagship store in Melbourne (discussed more in Section 1.5). Transnational initiatives and digital media mobilise new forms of solidarity and unlikely alliances across geography, class, race, gender and nationality beyond the immediate vicinity. Yet, for this 'politics of encounter' (Merrifield, 2013) to contribute to real democracy, more work is needed to unpack the process, complexities and local variations of urban assemblies to minimise exclusions. As transnational tech companies and computational infrastructure shape urban spaces, digital placemaking can no longer be contained within the sovereignty of one jurisdiction. Benjamin Bratton (2016) views software as a new layer of sovereign power that operates parallel to traditional forms of sovereignty, like nation-states and international organisations. Blok and Farías (2016) use urban cosmopolitics to describe the negotiation and contestation of urban assemblages—human and non-human elements that shape our cities. This approach acknowledges the ever-evolving nature of our common worlds, which are constantly 'being subtly transformed, destabilised, decentred, questioned, criticised or even destroyed' (Blok & Farías, 2016, p. 2).

Urban cosmopolitics also encourages digital placemaking studies to consider the nuances and diversity of non-Western approaches. Jennifer Robinson (2006) reminds us that Western urban experiences and organisations are not universally applicable models for city developments, nor should researchers deem a Western notion of modernity to be the sole origin of a linear history. Seeing urban innovations as results of global interdependence and learning, she argues that if cities are to be imagined on equitable footings, we need to develop post-colonial urban theories with resources as cosmopolitan as cities themselves. A cosmopolitical approach to digital placemaking studies prioritises actual situations and practical concerns as the driving forces and the source of the dynamics of the world. Acknowledging multiple and liquid modernity (Bauman, 2013) leads to re-historicising and reordering digital placemaking as a networked planetary co-existence.

The four horizons of mediated placemaking, assemblage urbanism, participatory condition and urban cosmopolitanism provide theoretical buffering to seven core challenges synthesised by Daniel Ross (in Stiegler, 2018, p. 11) as rise of social networks, ubiquitous screens, global financial crises, geopolitical crises, automation, Anthropocene's existential threat and an industrially generated populism. I further propose three pathways for this book to approach these horizons. First, as digital placemaking is seldom discussed beyond a Western liberal-democratic context, this book embeds the Asia-Pacific narrative into the global discourse and landscape of placemaking, focusing on examples from Australia, China and Taiwan. Aligning with Robinson's (2006) exploration of 'ordinary places' in 'ordinary cities', I delve into their complexity, diversity and distinctiveness. Rather than underscoring peculiarities, I celebrate the sheer variety of digital placemaking terminologies, origins, manifestations and governance philosophies towards the theory-building agenda of relational comparison (Söderström, 2014) and comparative urbanism.

Moreover, comprehensive publications specifically addressing the interplay between digital placemaking and urban politics are scarce. Early discussions treated digital placemaking as conceptual renderings, discrete applications or one-off projects. This book scrutinises a broad spectrum of real-world examples with the socio-technical milieu across three political systems to elucidate variations and the enduring impact of digital placemaking. It persistently revisits the Apple Federation Square case, dissecting its multifaceted aspects and ultimately reconstructing its repercussions, portraying digital placemaking as an emerging yet ongoing part of socio-political transformation.

Lastly, this book pioneers the theorisation of digital placemaking through the lens of urban assemblage. The assemblage approach probes into the specific instantiations of participation and political modes enacted at the confluence of places, technologies and practices. I explore different strands of socio-material urban peculiarities through what I term activist, organisational and governmental assemblages. By bringing the civic, business and government sectors into contextualised conversations, I trace how various urban segments, departments and institutions align and collide over time with considerations of urban nuance, relations and fluidity. This book examines digital placemaking beyond the place boundary, jurisdiction and sovereignty to reveal it as a continued (though overlooked) tension between global platformisation and urban governance. This approach challenges existing normative and hierarchical participatory models and a linear understanding of urban politics. It suggests a shift from, in Mol's (2002) terms, a 'politics of who' to a 'politics of what', which asks not only who should decide but also what should be done.

1.3 Existing Models of Urban Participation: Arnstein's Ladders and Beyond

Existing scholarship often empirically assesses urban participation as a measurable aspect of urban governance. Sherry R. Arnstein's (1975) 'ladder of public participation' ranks public participation forms into different hierarchies, visualising three categories (non-participation, tokenism and citizen power) across eight rungs (ranging from empty gestures to real empowerment). The apex, 'Citizen control', is achieved when citizens attain full managerial authority to develop and manage a place (p. 217). An experimental realisation of citizen control is in the form of a neighbourhood corporation, where communities foster economic enterprises and make decisions with government-allocated funds. The enduring influence and heuristic value of Arnstein's Ladder underpin many foundational concepts shaping public participation and practice globally.[7]

Paolo Cardullo and Rob Kitchin (2019, p. 11) extended Arnstein's framework to a conceptual scaffold of smart citizen's participation, reviving the citizen-led aspirations in urban governance against the technocratic and neoliberal orientations of smart cities. They inserted a new category, 'Consumerism', between 'Non-participation' and 'Tokenism' to denote citizens as consumers of urban services traditionally delivered by city administrators, now increasingly outsourced to private corporations or public–private partnerships. The scaffold enriches the original ladder with four additional columns: the roles of the citizens, forms of citizen involvement, justifying political discourse and nature of citizen participation.

These models are timely and significant. Arnstein's Ladder emerged in the 1960s, a decade marked by global social movements advocating citizens' right to the city (Lefebvre, 1996) and community agency (Polsby, 1980). Broadened participation and redistributed power were regarded as cornerstones of democracy, evidenced by experimental initiatives like the Model Cities Program (1966–1974) led by Arnstein herself. The scaffold applies to contemporary contexts, including large-scale regeneration through arts, cultural and community engagements, digital urban communication and the rise of neoliberal cities, marked by marketising urban governance and privatising public spaces (Brenner & Theodore, 2002).

However, normative models like these have limitations in accounting for diversity, contingency and complexity over time. First, both models rest on the presupposition that idealises full public participation on an imaginary echelon, implying a power hierarchy and absolute political opposition between the citizens and officials. They both assume democracy materialises when citizens achieve full autonomy, a notion historically debatable, as seen in theoretical examinations of the 'crowds' from Plato to Elias Canetti (e.g., McClelland, 2010). Participation acts as a pharmakon rather than an elixir. Scholars question whether direct participation or 'citizen control' is always desirable (e.g., Collins & Ison, 2009). The Victorian Auditor-General has recognised the merits and limitations of each level of the public participation and recommended local governments to tailor the use to achieve different aims rather than blindly pursuing the highest forms of citizen power under all circumstances. Similarly, inserting 'Consumerism' between 'Non-participation' and 'Tokenism' introduces a sense of absolutism. Modern consumers do more than choose to 'opt in' or 'out' of a service contract; they duly research, judge, compare, assess and publicly review offerings, aligning business decisions with political aptitudes (Zinnbauer, 2023). By undervaluing consumers, these models anticipate a self-fulfilling neoliberal cycle leading to citizen defeat. Ultimately, by arguing that proper forms of citizen participation should be included in the political, both models prioritise overt and temporary political struggle over the cumulative effects of quotidian actions in urban transformation.

Second, both models focus on binary relationships in placemaking politics, neglecting intermediaries that facilitate and catalyse change. Arnstein's (1975, p. 217) ladder aimed to heuristically 'highlight the fundamental divisions' between the powerless citizens and the powerful urban official. Similarly, Cardullo and Kitchin (2019) presupposed urban politics as a binary confrontation between citizens and neoliberal governments in the smart era. Yet the powerful and powerless are not clear-cut or stagnant categories. They are relative and fluid. In addition, the models do not acknowledge 'intermediaries' in assemblage urbanism nor concern entities that 'stand[ ] at a place in the network between two other actors and serve[ ] to translate between the two actors in such a way that their interaction can be more effectively coordinated' (Beveridge & Guy, 2011, p. 96). Many significant but somewhat neglected professions mediate placemaking politics, such as place managers, urban communications, artists, marketers, consultants and logistical administrators (Petersen, 2021). Since intermediaries can alter and stabilise relations between the regulator and the user, their contribution should be adequately accounted for in placemaking politics.

Lastly, both models engage with a limited range of carefully curated examples. Arnstein's ladder focuses on the 150 US cities selected for the Model Cities Program by the Lyndon Johnson Administration, reflecting 1960s social norms and political situations. The scaffold examines service-oriented digital apps and street furniture, possibly missing the broader spectrum of citizen power, such as urban activism or branding. This limited scope may lead to optimism in new activities like civic hackathons, underrating the liberating potential of essential urban services like traffic control. Urban activism, a crucial form of public participation, is omitted by both models. Contemporary urban activism and protests address and correspond with neoliberal designs and creative place branding, where counter-cultural activists' efforts are co-opted by neoliberal states and investors (Mayer, 2013). The democratic prospects of these activities require long-term validation rather than immediate celebration.

Unanswered questions persist: How can we reinvent a model to explore the politics of digital placemaking while accommodating heterogeneity and interconnection? How can we break free from a rigid, hierarchical political imagination and recognise a mode of politics where heterogeneous actors emerge and converge in longitudinal political processes? In other words, how do we reimagine the politics of digital cities without assuming, vilifying or idealising an ultimate reality but accept the world as a collision of different processes, realities and interpretations? All of us are part of the world and responsible for shaping it towards an agreed-upon future.

1.4 The Ouroboros of Digital Placemaking

An assemblage-based model of urban cosmopolitics, which I term the ouroboros of digital placemaking, can counteract political myopia and determinism by restoring a focus on complexity, emergence and multidirectional causality rather than emphasising one universal set of laws as the fundamental way to comprehend digital cities. For Isabelle Stengers (2005, p. 995), cosmopolitics is 'the unknown constituted by these multiple, divergent worlds, and [] the articulations of which they could eventually be capable', a philosophy that gives space for the emergence. The political challenge in addressing this unknown complexity is to embrace our agnosticism with humility to acknowledge that current ethics and morals may be inadequate for emerging priorities, obligations and requirements. It necessitates critical reflection on proposed solutions that envision an improved future consisting solely of 'good cosmopolitans' and compelling others to conform. It involves normalising conflicts as inherent to cosmological pluralism rather than demonising opponents.

When reflecting on Stengers's (2005) notion of 'temporary cessation,' imagine an amber pendant encapsulating a moment in time. Visualise a spider, thousands of years ago, raising its body and opening its fangs to confront an ensnared moth. The moth, attempting to escape, readied itself for an attack. Suddenly, a drop of pine resin fell, capturing the pair and fossilising their struggle perpetually. Revisiting this moment, we could critique the structural and biopolitical inequalities or lament the predator/prey disparity. However, such 'who politics' can become cliché. What is more productive is 'how politics.' We need to understand how the moth could find a line of flight. What allies could it strategically mobilise to achieve liberation? Perhaps a spiderweb, a twig, other spiders—there may have been a movement to make or an angle of retreat. The responsibility of urban cosmopolitics is to find that line of flight *in situ* and restore hope. It offers hope for those who recognise when to harness the power that permeates the assemblages, rather than succumbing to learned helplessness. Such is the promise of an assemblage-based cosmopolitical model.

The ouroboros of digital placemaking (Figure 1.1) represents an iterative process that potentiates multimodal and assemblage-based political reflexivity. An ouroboros, a snake biting its own tail, embodies both generative and destructive forces. This concept draws from Latour's political circle (Latour, 2007, 2013), condensing his five political stages into three interweaving assemblages: activist, organisational and governmental. The trajectory of an issue follows emergence, turbulence, institutionalisation, solution and routinisation. Politics in digital cities emerge as new human/non-human associations form in political-1. Public discontent is articulated and performed in the activist assemblage in political-2. The governmental assemblage transforms the contested issue into a public-facing question addressed through established institutions and policy frameworks in political-3. All assemblages mobilise various forms of public consultation and deliberation to deliver solutions in political-4. The organisational assemblage incorporates the solution and routinises it in daily administration in political-5. Differing from a circular model that presumes restoration of the former status, the ouroboros illustrates an iterative trajectory with potential for reflexivity, rebirth or demise, tracing urban potentiality rather than claiming certainty or an 'ultimate reality'. In other words, rather than implying *da capo*, or going back to the beginning, the ouroboros demonstrates a thesis of constant change.

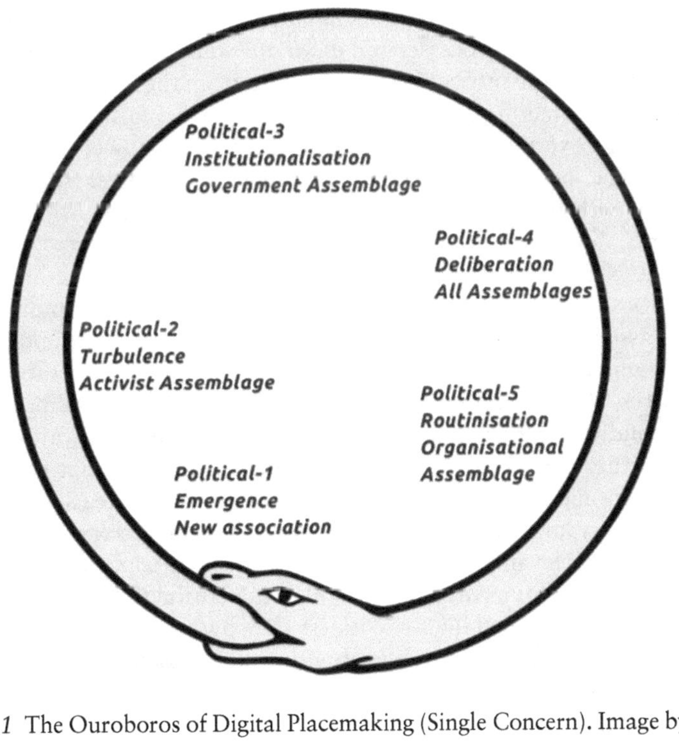

Figure 1.1 The Ouroboros of Digital Placemaking (Single Concern). Image by author.

The framework gives theoretical symmetry to participatory power in urban politics rather than treating it as an afterthought or a concession of fragmented and paradoxical urban realities.

This model addresses the four horizons for digital placemaking studies and reimagines urban politics. First, the ouroboros applies sophistication in unpacking its heterogeneous actors and fissured processes. A video art installation in a city mall exemplifies this intersection, functioning at the confluence of video art, commercial television and surveillance practices, where the 'screen functioned like a dynamic, expansive frame with moving times and spaces of its own' (Verstraete, 2016, p. 312). Contemporary public spaces are networked environments filled with interrelated entities, affect and associations that transform, distort and multiply time and space, producing their own realities while responding to local and global forces (Low, 2023; Papastergiadis, 2016). New socio-technical associations wield power to destabilise and confront society with unprecedented controversies. For Latour (1984), democracy begins as an inquiry into a disturbed *res publica*, where 'a thing emerges before anything else as a scandal at the heart of an assembly that carries on a discussion requiring a judgment brought in common.' Political-1 in the ouroboros is sensitive to emerging disruptions and surprises unveiled by new socio-technical practices.

The ouroboros also comprehends contemporary urban politics at the intersection of activism, economy and governance. The sense of politics as absolute opposition between irreconcilable enemies is renewed as the clash of concerns, perspectives, processes and *modi operandi*. Political-2 of the ouroboros demonstrates the intricate politics arising from the fluidity and synergies of different placemaking assemblages. This model reveals how the prescriptive power of governmental institutions, operational power of place industries, agonistic power of activists and generative power of the publics experiment with new collaboration methods. Instead of a 'winner-takes-it-all' scenario, different parties respond by translating and effectively betraying one another's literal wishes (Harman, 2014). The assemblage-based model accounts for internal tension and fluidity in each political step and transitions from one assemblage to another. The ouroboros extends urban politics from the embryonic moment of conflict into urban routines and operations.

Moreover, the ouroboros realistically reflects public participation as an ambient condition rather than idealising it as a sublime political outcome or absolute virtue. Latour (2007) controversially maintained that participatory politics is optional with selected agenda items. Despite increasing digitisation and connectivity, and politics becoming more planetary, territorial proximity and hyper-local relevance still warrant real-world participation. For instance, it makes little sense for a rural Queenslander to petition the City of Melbourne (COM) against a planned high-rise unless they own nearby real estate. Issues vary in their ability to retain public attention and attract participation in different forms. Those who protested for democracy in Hong Kong cannot rightfully blame those who stayed home, as the latter may have other political priorities (i.e., belonging naturally to other political catchments) or be entirely disengaged. The ouroboros is empathetic yet not requiring the public to be omnipotent and intervene in every issue whenever possible. It acknowledges that interested publics may change at every step, allowing time and space for different participatory modes such as observation, activism, utilisation, consumption and voting. Thus, what was previously conceived as specific rungs of the ladder and scaffold become multiple entry points into the ouroboros, elucidating participatory possibilities in different stages of urban politics.

Finally, the ouroboros is context-sensitive and place-based, applicable to all societies with activist, organisational and governmental assemblages. An assemblage incorporates heterogeneous and interconnected components not traditionally considered together. Rather than emphasising positions, the ouroboros highlights relations and their actualisation in different contexts. For example, the activist assemblage differs from an activist group, lacking strict group boundaries and including all 'activist-minded' supporters. Assemblage thinking positions the activist assemblage in relation to non-activists such as media professionals, the general public, organisational and governmental assemblages and supporting complex urban milieu. Relationality and contextuality prevent assemblages from being predictable and replicable. The ouroboros avoids reducing non-Western cities to a synecdoche of their Western counterparts or vice versa. As demonstrated in this book, the ouroboros occurs in both democracies and non-democracies. Contrary to conventional wisdom asserting minimal public participation in non-democracies, the ouroboros highlights the resilience of activist assemblages and pervasive modes of public participation even in non-democracies. However, democratic processes accommodate the free movement of issues followed by legit and accountable parties through a complete cycle, while non-democracies tend to contain the issues within one political step and prevent them from progressing to the next steps.

1.5 Contextualising Federation Square

A mundane episode in December 2017 triggered my systematic reflection on digital placemaking. Apple Inc., endorsed by the Victorian government, announced a plan to open a flagship store on Federation Square, a high-profile public space in Melbourne where I live. The 'Apple Federation Square' proposal included photography, music, gaming and app development workshops alongside retail activities. Despite endorsements from Federation Square's management and the Victorian Government, the decision faced significant public discontent, evolving into online petitions, public discussions and urban activism. Consequently, Apple's demolition request was rejected by the Victorian Heritage Authority, and the plan was shelved following the resignation of Angela Ahrendts, the mastermind behind Apple's 'town square' initiative. This led to the Victorian Government launching the Federation Square Review to examine its financial health and governance and an urban hackathon to co-design its future.

Federation Square, commissioned by the Victorian government to mark the Australian Federation's centenary at the dawn of the 21st century, filled a 200-year void in Melbourne's urban landscape—long devoid of public squares due to colonial fears of mass gatherings and rebellion. When it was officially opened in 2002, its postmodern aesthetics, angular design and sloping topography elicited mixed public reactions amidst controversies of budget overruns and delayed delivery (Brennan, 2010). Embracing the concept of a 'federation,' the design unified disparate structures into a cohesive whole, mirroring post-1960s trends of less overtly patriotic public spaces, influenced by European open-space paradigms (Frost, 2012). According to its principal architects, Federation Square aimed to create a dynamic urban order adaptable to social changes, maintaining spatial fluidity and semi-permeability (Bates & Davidson, 1999).

Spanning 5.2 hectares, Federation Square integrates cultural, civic and commercial elements, hosting over 40 cultural institutions, including the National Gallery Victorian Australian Collection, the Australian Centre of Moving Images, the Special Broadcasting Service, the Koorie Heritage Trust and the Melbourne Festival Headquarters. Its central arena, with a semi-permanent stage and an amphitheatre-like open space, features a large urban digital screen, the Digital Façade, launched in 2018. This A$5.4-million ($3.6-million) installation, comprising over 850 light-emitting diode (LED) panels, showcases artworks, broadcasts sports and hosts interactive games. The paved deck, extending to the Yarra River and Birrarung Marr, is surfaced with distinct sandstone from the Kimberly region and features designer Paul Carter's artwork, which embeds stories from Australia's placemaking history (Carter, 2005). Commercial amenities include restaurants, cafes, bars, stores and a 500-car parking facility.

In 1999, the Victorian Government established Fed Square Proprietary Limited (FSPL) to manage Federation Square, operating as a not-for-profit custodian under Victorian State Trustees Limited. FSPL's mandate includes tenancy coordination, public space programming, marketing, asset management and development, reporting to its Board of Directors and functioning with a dedicated team of 40 to 50 staff. Initially, the COM planned to co-fund the project, but cost overruns led to its withdrawal (Brennan, Interview, 14 May 2018), rendering Federation Square a predominantly state project. FSPL's policy, the Civic and Cultural Charter, guides the balance of Federation Square's civic, cultural and commercial functions, dictating its operations, marketing and event programming. The charter stipulates that commercial uses of the space should have an identifiable synergy with Federation Square's cultural program. Yet its vagueness on budgetary, financial and maintenance specifics left operational balance ambiguous.

As a public non-financial corporation, FSPL operates independently of recurring government funds, relying on tenancy rents, parking fees, commercial events, sponsorships, and grants to cover expenses and infrastructural maintenance. Its management and placemaking strategies have evolved significantly under eight CEOs over two decades. Based on interviews and annual reports, FSPL's philosophy has transitioned through four stages: public infrastructure management (2002–2005), events and cultural programming (2006–2015), commercial planning (2015–Aug. 2020) and the Victorian creative institution period (Aug. 2020–present). In August 2020, it became one of nine state-owned creative organisations under Creative Victoria, following participatory events and political developments.

Apple Federation Square illustrated the diverse *modi operandi* of digital placemaking in contemporary cities. It began with Apple's techno-capitalist proposal to alter a public square's spatial arrangement, triggering urban activism that adopted hybrid techniques—a design hackathon embodying the digital logic of user-generation and a state review supported by government engagement portals. This case provided a prototype of Australia's digital placemaking, which I later applied to cases in China, Taiwan and beyond. This prototype informs much of discussion in this book, to which I will return throughout.

The demise of Apple Federation Square suggested that digital city politics are multilateral. Simple binaries of top-down technocratic and bottom-up community orientations do not suffice, nor do absolute notions of winning or losing. The long-term effects of digital placemaking are still unfolding, and it remains ambiguous whether Apple Federation Square was a triumphant public intervention or a missed opportunity. This book seeks to understand intricate urban dynamics and identify a sense of urban politics extending from moments of conflict into urban routines and operations. I propose several new concepts, including the ouroboros of digital placemaking, activist, organisational and governmental assemblages and socio-technical natality. The ouroboros conceptualises urban politics as a practice involving humans and artefacts. The three assemblages provide frameworks for understanding these interactions. Renewed comprehension of the Arendtian concept of natality depicts the possibility for human and non-human actants to collectively create new beginnings in a non-determinist and non-defeatist future.

1.6 Research Design

To trace the participatory and political dynamics of digital placemaking, I analyse a single-site case study through ethnography informed by the Actor-Network Theory (ANT). Specifically, I examine Federation Square and three interconnected micro-cases: Our City, Our Square; Hack Fed Square; and Federation Square Review. These micro-cases reside in three digital placemaking assemblages: activist assemblage (Chapter 3), organisational assemblage (Chapter 4) and governmental assemblage (Chapter 5). Cases from other political contexts are also brought into the discussion of the assemblages to test the applicability of the ouroboros model.

In an ANT-informed account, the 'site' functions as a mediating interface between architecture, nature and society, caught between objective reality and subjective perception or social relation, thus emerging as a set of movements to follow (Yaneva and Mommersteeg, 2019, p. 306). Melbourne's Federation Square illustrates the processual and evolving nature of urban socio-technical assemblages. It is not only a pioneering institution for urban cultural experimentation and an early adopter of digital placemaking but also a known battlefield for placemaking conflicts, exemplified by the Apple controversy (extrapolated throughout Chapters 2–5). This process aligns with what Latour (2005, p. 87) termed 'controversy identification,' marking the research project's inception when the researcher senses or observes something unusual and is compelled to explore the issue.

ANT-informed ethnography differs from the 'classical' ethnographic approaches. Classical Malinowskian ethnography describes studying an exotic and distant 'other' who is different from the researchers themselves (Malinowski, 1922). This notion has since then been critiqued and updated to reflect a blurred understanding of the site boundaries (Marcus, 1995) and a less hegemonic approach that assumes the researchers' superiority and objectivity (Gupta & Ferguson, 1997). ANT-informed ethnography, rejecting the prioritisation of humans over non-humans, adopts a socio-material stance (MacLeod et al., 2019), moving beyond traditional ethnographic focus on human interactions to symmetrically address human and non-human elements (Fenwick & Nimmo, 2015). This approach accommodates data collection across the socio-material web of relationships in digital placemaking. These include material elements like architectural and media infrastructure and human bodies; discursive elements like plans, policies, strategies and public opinion; and affective elements like public experience and sentiment. This method addresses the contingency and heterogeneity of urban socio-technical assemblages through dynamic documentation of processes, negotiations and translations. Additionally, ANT-informed ethnography allows for the exploration of the emergent notion of digital placemaking, which is vaguely defined and agreed upon in practice. Other methodologies like surveys or structured interviews are less suitable, as they require predetermining the research problem and terminology, which is not ideal for exploratory research where terminology from different industries is unknown and in-depth information is needed to form the core research question.

My ANT-informed ethnography consists of participatory observation, semi-structured expert interviews and document analysis. Rather than assuming a purely observational role, I actively 'follow the actors themselves,' as Latour suggested (2005, p. 12), accepting the researcher as part of the assemblage in flux. Immersed in the socio-technical assemblage, I traced the actors and let them guide me to new events and encounters. I actively participated in events, rallies, consultations, hearings, meetings and conferences related to Federation Square. Many events were open to the public, while some opportunities arose by chance. I set up Google alerts, subscribed to email lists and frequently visited activist websites and Instagram accounts. The activist campaign became my first micro-case, involving major events like protests, pop-up street theatre and public rallies. I attended forums and talks with architects, planners, placemakers and community engagement consultants, including the Victorian Heritage Council's public hearing. I filed an expression of interest and was selected to participate in Hack Fed Square, which became my second micro-case. Subsequently, I was approached by the community engagement consultancy Capire to participate in the Federation Square Review, forming my third micro-case. Correspondence with local placemaking agencies, such as CoDesign, Neighbourlytics, Place Scores and VillageWell, and events held by CoDesign, led to further interactions with senior urban planners, engagement consultants and place managers from local councils.

This extensive engagement generated numerous photographs, audio and video recordings, event timelines, relationship maps, fieldnotes, new encounters, email introductions, expert interviews and informal conversations with practitioners in related fields. These interactions included engagement consultants, placemakers, artists, urban designers, council staff and Apple store experience leaders known as 'Apple geniuses,' providing triangulated information for this thesis. I also drew on documents such as government policies, Federation Square policies, annual reports from 2012 to 2020, media releases, the artist catalogue *Mythform* (Carter, 2005), activist archives, Hack Fed Square materials and Federation Square Review material. Additionally, I collected over 140 news articles to map the Apple controversy timeline and followed online discussions, petitions on public forums and relevant radio program transcripts.

1.7 Book Overview

This book examines the political dimensions of digital placemaking, proposing a tripartite analytical model to investigate its complexities. Chapter 2 situates digital placemaking within broader historical transformation and interdisciplinary debates, focusing on Australia, China and Taiwan. It critiques existing discourse that simplifies urban participation into binary oppositions between citizens and governments, advocating instead for an assemblage perspective to reveal digital placemaking as a relational site of contention and collaboration. I begin by surveying the diverse and emerging field of digital placemaking, highlighting its heterogeneous actors, practices and protocols. I argue that digital placemaking blurs the formal–informal binary and should be recognized as an interface of urban politics. Chapter 6 concludes by refining the notion of digital placemaking and discussing an agnostic mode of cosmopolitics. Recognising the multitude of practices and contexts, I contend that digital placemaking is an experimental field with socio-technical natality for urban politics beyond immediacy and locality.

Chapters 3–5 delve into what I term the activist, organisational and governmental assemblages. Chapter 3 examines the activist assemblage, exemplified by the 'Our City, Our Square' campaign (Melbourne, Australia) and 'Everybody's Donghu' campaign (Wuhan, China). These campaigns deploy sophisticated and hybrid techniques, distinguishing between performative and design interventions, highlighting socio-technical skills pervasive to the activist assemblages. Chapter 4 explores the organisation assemblage, exemplified by 'Hack Fed Square' (Melbourne) and Wuhan's *wanghong* (online celebrity) city. The chapter argues that business-led placemaking is driven by networked user-generated logic, site-specific labour and neoliberal market orientations, embedding commerce and community interest within urban complexities. Chapter 5 investigates the governmental assemblage, with examples like the Federation Square Review (Melbourne) and the g0v initiative (Taiwan). It considers these initiatives as collaborations among diverse participants capable of enacting structural changes. The governmental assemblage, viewed not as a monolithic entity but as a network with blurred boundaries and internal tensions, challenges the binary understanding of urban politics as confrontations between the 'have-nots' and 'powerholders.'

My transdisciplinary approach reflects a rejection of compartmentalising digital placemaking's fluid realities within rigid and abstracted disciplinary boundaries. By 'trespassing' (in Hirschman's term, 1981) into the transdisciplinary realm, I evoke digital placemaking as a site of urban struggles infused with latent and explicit political tension. In this dynamic space, urban planners encounter media architecture designers, media scholars engage with digital marketers, app designers, public officials and placemaking agencies. The book's title salutes Latour's 'Reassembling the Social,' prioritising the study of heterogeneous elements traditionally considered separately and addresses power relations by reassembling the tension and the vitality of digital placemaking at the intersection of local happenings, global trends, democratic advancements and neoliberal logic.[8]

The cases in this book are meant to be not comprehensive but illustrative. They serve as an oligopticon, a concept by Bruno Latour (2005, p. 181), a special site where the micro-structures of macro-phenomena are crafted. Unlike the panopticon', which imposes total surveillance, the oligopticon provides a robust but narrow view to understand the connected whole through focused lenses (Latour & Hermant, 1998). It captures the cosmopolitics of commoning, representing an amalgam of multiple scales, temporalities, visibilities and representations. Rather than focusing on specific activities or actor groups, I trace the dynamics and interrelations among different assemblages that energise those activities.

This book addresses a crucial concern for contemporary urban governance: the integration of the digital, urban and political. Digital technologies shape urban politics while generating controversy. By addressing theoretical and practical aspects of digital placemaking, I respond to Gibson-Graham's (2006): If we recognise emerging technologies as inherently political and entangled in power-laden socio-technical assemblages, then how can we reimagine the postcapitalist politics of digital technologies? Urban governments should harness the technological moment known as the digital transformation towards cosmopolitics and democracy. I argue that digital placemaking provides a new testing ground for urban democracy, providing a relational site of contentions and collaborations among activists, professionals and governments, transcending strict boundaries and reimagining urban politics amidst complexities.

Notes

1 As the national legislature of the People's Republic of China, the National People's Congress consisted of 2,980 delegates indirectly selected by provincial governments in 2018.

2 See Chinese New Smart City Evaluation Indicator (GB/T 33356-2016) [in Mandarin], Section 3.3.2 Public Safety.

3 As is evidenced by the 2024 US Presidential Election, where tech mogul Elon Musk was promised a lead position the newly formed Department of Government Efficiency.

4 While the spelling variant 'place-making' predominates in extant literature (Basaraba, 2021), this book adopts 'placemaking' and 'digital placemaking' to align with the broader trajectory of media studies scholarship. This choice does not endorse a constructed dichotomy between 'organic place-making' and 'planned placemaking' (see Lew, 2017, p. 448).

5 The urban as a 'multiple object' has been inspired by Mol's (2002) conception of 'multiple bodies' that is ontologically in-between a totality and fragment, singular and plural.

6 Stiegler (2010) borrowed the concept from Derrida's reading of Plato.

7 See a list of 60 international public participation models summarised by Granicus, https://granicus.com/blog/international-public-participation-models/. These include the Community Engagement Model (2014) by the International Association of Public Participation (IAP2), a professional engagement body recognised widely in Australasia.

8 For the limitations of assemblage urbanism, see the debates between assemblage urbanists and critical urbanists in the special issue (2009) of *City: analysis of urban trends, culture, theory, policy, action* and related articles (2011). Assemblage urbanism was predominantly criticised for lacking focus on power relations and ignoring the 'context of context', i.e., capitalism (see Brenner et al., 2011).

References

Arnstein, S. (1975). Whom does the advocate planner serve (part two): Additional commentaries. In R. Cloward & F. F. Piven (Eds.), *The politics of turmoil: Poverty, race and the urban crisis* (pp. 54–55). New York: Vintage Press.

Barney, D., Coleman, G., Ross, C., Sterne, J., & Tembeck, T. (Eds.). (2016). *The participatory condition in the digital age* (Vol. 51). Minneapolis: University of Minnesota Press.

Barns, S. (2020). *Platform urbanism: Negotiating platform ecosystems in connected cities*. Singapore: Springer.

Basaraba, N. (2021). The emergence of creative and digital place-making: A scoping review across disciplines. *New Media & Society*, *25*(6), 1470–1497.

Bates, D. L., & Davidson, P. (1999). Federation Square, Melbourne, Australia Lab Architecture Studio. *Assemblage*, 40, 57–67.

Bauman, Z. (2013). *Liquid modernity*. New York: John Wiley & Sons.

Beveridge, R., & Guy, S. (2011). Innovation to intermediaries: Translating the EU Urban wastewater directive. In S. Guy, S. Marvin, W. Medd, & T. Moss (Eds.), *Shaping urban infrastructures* (pp. 92–107). Earthscan.
Blok, A., & Farías, I. (Eds.). (2016). *Urban cosmopolitics: Agencements, assemblies, atmospheres*. London and New York: Routledge.
Bratton, B. H. (2016). *The stack: On software and sovereignty*. Cambridge, MA: MIT Press.
Brennan, K. (2010). Federation Square–a place for the people. *Journal of Place Management and Development*, 3(2), 130–135.
Brenner, N., Madden, D. J., & Wachsmuth, D. (2011). Assemblage urbanism and the challenges of critical urban theory. *City*, *15*(2), 225–240.
Brenner, N., & Theodore, N. (2002). Cities and the geographies of "actually existing neoliberalism". *Antipode*, *34*(3), 349–379.
Cardullo, P., & Kitchin, R. (2019). Being a 'citizen' in the smart city: Up and down the scaffold of smart citizen participation in Dublin, Ireland. *GeoJournal*, *84*(1), 1–13.
Carter, P. (2005). *Mythform: The making of Nearamnew at Federation Square*. Melbourne: Miegunyah Press.
Collins, K., & Ison, R. (2009). Jumping off Arnstein's ladder: Social learning as a new policy paradigm for climate change adaptation. *Environmental Policy and Governance*, *19*(6), 358–373.
Cresswell, T. (1992). *In place-out of place: Geography, ideology, and transgression* (Vol. 2). Minneapolis: University of Minnesota Press.
De Munck, B. (2017). Re-assembling actor-network theory and urban history. *Urban History*, *44*(1), 111–122.
Deleuze, G. & Guattari, F. (1987). *A thousand plateaus: Capitalism and schizophrenia*. Bloomsbury Publishing. B. Massumi (tran). Minneapolis, London: University of Minnesota Press.
Farías, I. & Bender, T. (eds.) (2010) *Urban assemblages: How actor-network theory changes urban studies*. London: Routledge.
Fenwick, T., & Nimmo, G. (2015). Making visible what matters: Sociomaterial approaches for research and practice in healthcare education. In J. A. Cleland & S. J. Durning (Eds.), *Researching medical education* (pp. 67–79). New York: John Wiley and Sons.
Foth, M., Brynskov, M., & Ojala, T. (2015). *Citizen's right to the digital city*. Berlin: Springer.
Fredericks, J., Hespanhol, L., & Tomitsch, M. (2016). Not just pretty lights: Using digital technologies to inform city making. In *Proceedings of the 3rd Conference on Media Architecture Biennale* (p. 7). ACM.
Frost, W. (2012). Commemorative events and heritage in former capitals: a case study of Melbourne. In: Maitland, R. (Ed.). (2014). *Tourism in national capitals and global change*. Routledge. pp. 48–57.
Gibson-Graham, J. K. (2006). *A postcapitalist politics*. Minneapolis: University of Minnesota Press.
Graham, M. (2020). Regulate, replicate, and resist–the conjunctural geographies of platform urbanism. *Urban Geography*, *41*(3), 453–457.
Gupta, A., & Ferguson, J. (1997). Discipline and practice: 'The field' as site, method, and location In: A. Gupta and J. Ferguson (eds), *Anthropological locations: Boundaries and grounds of a field science*. Berkeley, Los Angeles, London: University of California Press, pp. 1–46.
Habermas, J. (1991). *The structural transformation of the public sphere: An inquiry into a category of bourgeois society*. Cambridge, MA: MIT Press.
Harman, G. (2014). *Bruno Latour: Reassembling the political*. London: Pluto Press.

Hes, D., Mateo-Babiano, I., & Lee, G. (2020). Fundamentals of placemaking for the built environment: An introduction. In: D. Hes and C. Hernandez-Santin (eds), *Placemaking fundamentals for the built environment* (pp. 1–13). Singapore: Palgrave Macmillan.

Hespanhol, L., Haeusler, M. H., Tomitsch, M., & Tscherteu, G. (2017). *Media architecture compendium: Digital placemaking.* Stuttgart: Av edition.

Hirschman, A. O. (1981). *Essays in trespassing: Economics to politics and beyond.* Cambridge, MA: Cambridge University Press.

Kellner, D. (2021). *Technology and democracy: Toward a critical theory of digital technologies, technopolitics, and technocapitalism.* Springer Nature.

Kitchin, R. (2014). The real-time city? Big data and smart urbanism. *GeoJournal*, *79*(1), 1–14.

Kitchin, R. (2015). Making sense of smart cities: addressing present shortcomings. *Cambridge Journal of Regions, Economy and Society*, *8*(1), 131–136.

Latorre, D. (2011, September 21). *Digital Placemaking – Authentic Civic Engagement.* Project for Public Spaces. https://www.pps.org/article/digital-placemaking-authentic-civic-engagement

Latour, B. (1984). The powers of association. *The Sociological Review*, *32*, 264–280.

Latour, B. (2005). *Reassembling the social: An introduction to actor-network-theory.* Oxford, New York: Oxford University Press.

Latour, B. (2007). Turning around politics: A note on Gerard de Vries' paper. *Social Studies of Science*, *37*(5), 811–820.

Latour, B. (2013). *An inquiry into modes of existence.* Harvard University Press.

Latour, B., & Hermant, E. (1998). *Paris: Ville Invisible/Paris: invisible city.* Paris: Éditions la Découverte.

Lefebvre, H. (1970/2003). *The urban revolution.* Minneapolis: University of Minnesota Press.

Lefebvre, H. (1996) *Writings on cities.* Oxford: Blackwell Publishers Ltd.

Leszczynski, A. (2020). Glitchy vignettes of platform urbanism. *Environment and Planning D: Society and Space*, *38*(2), 189–208.

Lew, A. A. (2017). Tourism planning and place making: place-making or placemaking? *Tourism Geographies*, *19*(3), 448–466.

Low, S. M. (2023). *Why public space matters.* Oxford: Oxford University Press.

Lynch, C. R. (2020). Contesting digital futures: Urban politics. Alternative Economies, and the Movement for Technological Sovereignty in Barcelona. *Antipode*, *52*(3), 1–21.

MacLeod, A., Cameron, P., Ajjawi, R., Kits, O., & Tummons, J. (2019). Actor-network theory and ethnography: Sociomaterial approaches to researching medical education. *Perspectives on Medical Education*, *8*(3), 177–186.

Malinowski, B. (1922). *Argonauts of the Western Pacific: An account of native enterprise and adventure in the Archipelagoes of Melanesian New Guinea.* New York: EP Dutton & Co.

Marcus, G. E. (1995). Ethnography in/of the world system: The emergence of multi-sited ethnography. *Annual Review of Anthropology*, *24*(1), 95–117.

Mattern, S. (2017). *Code and clay, data and dirt: Five thousand years of urban media.* University of Minnesota Press.

Mayer, M. (2013). First world urban activism: Beyond austerity urbanism and creative city politics. *City*, *17*(1), 5–19.

McClelland, J. S. (2010). *The Crowd and the Mob (Routledge Revivals): From Plato to Canetti.* Routledge.

McFarlane, C. (2011). Assemblage and critical urbanism. *City*, *15*(2), 204–224.

McQuire, S. (2017). *Geomedia: Networked cities and the future of public space.* Hoboken, NJ: John Wiley & Sons.

Meadows, D. H., Meadows, D. L., Randers, J., & Behrens III, W. W. (1972). *The limits to growth; A report for the club of Rome's Project on the Predicament of mankind*. New York: Universe Books.
Merrifield, A. (2013). *The politics of the encounter: Urban theory and protest under planetary urbanization* (Vol. 19). Athens: University of Georgia Press.
Mol, A. (2002). *The body multiple: Ontology in medical practice*. Durham: Duke University Press.
Mumford, L. (1938). *The culture of cities*. New York: Harcourt, Brace and Company
Offenhuber, D., & Ratti, C. (2014). *Decoding the city: Urbanism in the age of big data*. Basel: Birkhäuser.
Osborne, S.P. (eds.) (2005). *The new public governance?: Emerging Perspectives on the theory and practice of public governance*, London and New York: Routledge.
Papastergiadis, N. (Ed.). (2016). *Ambient screens and transnational public spaces*. Hong Kong University Press.
Petersen, L. K. (2021). Caretakers as mediators in household waste separation. *Journal of Environmental Policy & Planning*, 23(6), 1–13.
Polsby, N. W. (1980). *Community power and political theory: A further look at problems of evidence and inference*. 2nd ed. New Haven, CT: Yale University Press.
Polson, E. (2015). A gateway to the global city: Mobile place-making practices by expats. *New Media & Society*, *17*(4), 629–645.
Robinson, J. (2006). *Ordinary cities: between modernity and development*. London: Psychology Press.
Sadowski, J. (2021). Who owns the future city? Phases of technological urbanism and shifts in sovereignty. *Urban Studies*, *58*(8), 1732–1744.
Schneekloth, L. H., & Shibley, R. G. (1995). *Placemaking: The art and practice of building communities*. New York: Wiley.
Sloterdijk, P. (2016). *Foams: Spheres volume III: Plural Spherology*. Los Angeles: Semiotext(e).
Söderström, O. 2014. *Cities in relations: Trajectories of urban development in Hanoi and Ouagadougou*. New York: John Wiley & Sons Inc.
Stengers, I. (2005). The Cosmopolitical Proposal. In *Making things public: Atmospheres of democracy*. Bruno Latour and Peter Weibel, eds. Pp. 994–1004. Cambridge, MA: MIT Press.
Stiegler, B. (2010). *For a new critique of political economy*. Hoboken, NJ: John Wiley.
Stiegler, B. (2018). *The neganthropocene* (p. 349). Open Humanities Press.
Tomitsch, M. (2016). Communities, spectacles and infrastructures: Three approaches to digital placemaking. In S. Pop, T. Toft, N. Calvillo, & M. Wright (Eds.), *What urban media art can do: Why, when, where, and how*. Stuttgart: Av edition.
Tuan, Y. F. (1975). Place: An experiential perspective. *Geographical Review*, 65(2), 151–165.
Varoufakis, Y. (2024). *Technofeudalism: What killed capitalism*. Melbourne: Melville House.
Verstraete, G. (2016). Media, places, and assemblies: From videowalls to "smart" environments. *Space and Culture*, *19*(4), 308–318.
Yaneva, A., & Mommersteeg, B. (2019). How does an ANT approach help us rethink the notion of site? In: A. Blok, I. Farias, & C. Roberts (Eds.), *The Routledge companion to actor-network theory* (1st ed., pp. 306–317). London and New York: Routledge.
Zinnbauer, D. (2023). Full-stack Citizenship 2028. http://doi.org/10.2139/ssrn.4598061

2 Theorising Digital Placemaking

2.1 Placemaking: A Fraught Chronology of Public Participation

What does digital placemaking actually entail? A cursory Google search for 'digital placemaking' yields millions of results, revealing a plethora of spellings like 'digital place-making' and related practices not explicitly categorised as placemaking. Among these results are promotional websites, project showcases by urban designers, public media artists, consultancies, engagement and tourism agencies, local governments and an expanding corpus of academic research. Digital placemaking is an industry-driven field with a burgeoning number of stakeholders, each vying to define the term according to their perspectives and interests, yet an authoritative academic definition remains elusive. This section endeavours to elucidate this extensive dataset by situating digital placemaking within broader historical contexts and diverse practices, commencing with an interdisciplinary review of place and placemaking.

In a broader sense, placemaking encompasses both organised and organic forms of human settlement that ceaselessly interact with the juridical, political, economic and cultural facets of society. Early placemaking tactics, such as zoning, planning, building codes and inhabitation, were essential in spatialising socio-political order and manifesting lived experience. Responding to what Christian Norberg-Schulz (1979) termed the *genius loci*, placemaking integrates the physicality of space with local culture, beliefs and spirituality. Classic European cities, functioning as self-organised local political systems through collective decision-making processes, often coalesced around a central public space for communal gathering, offering accessibility and visibility as the 'space of appearance' for pluralistic co-existence (Arendt, 1958). Conversely, traditional Chinese cities centred on the *yamen* (offices of local authorities) and *chenghuangmiao* (temple of the City God) as political and spiritual anchors for the population (Friedmann, 2007). Early spaces of civic freedom bourgeoned in semi-public and commercial settings like teahouse and side streets (Wang, 2003). Although the teahouse culture[1] came under strict regulations during the Republic of China in the late 19th century and disappeared during the Mao Zedong era, pubs and cafés developed into the Habermasian public sphere with the rise of the printing press.

DOI: 10.4324/9781003407973-2

In a narrower sense, placemaking eventuated as professional and commercial practices in the United States during the 1950s. Rather than as streamlined procedures and industry standards, placemaking materialised at the nexus of multiple actors, industries and historical trajectories. Since then, the rhetoric of placemaking has expanded to incorporate economic metrics, policy instruments and communications tools. This section provides parallel narratives of placemaking as a movement of ideas and practices reflecting global trends and local variations.

2.1.1 Placemaking and Placemaking Industries (1950s–1990s)

In the late nineteenth century, the nascent profession of urban planning aspired to impose order and structure to the unruly industrial cities through city-wide master plans encompassing sanitation, goods provision and transportation. However, a reductive tendency towards urbanisation risks producing homogenised urban landscapes. At its extreme, master planning's obsessive drive for monotonous and rational land use and infrastructural development eradicated local distinctiveness, resulting in standardised, austere and sterile spaces. Amid the structural economic and social changes that strained traditional spatial and urban relationships, scholars lamented a 'loss of place' or a sense of 'placelessness' (Relph, 1976), generating 'abstract space' (Lefebvre, 1991) and 'non-place' (Augé, 1995), further claiming to restore a sense of priority, 'first life, then space, then buildings' (Gehl, 2013). Such laments underscored the urgency of reinvigorating urban public spaces with meaning, depth and community experience, while catalysing the emergence of placemaking practices in the 1950s.

Placemakers criticised urban modernism for its priority on automobiles, infrastructures and capital at the expense of pedestrians, community and environment (Aravot, 2002). They also condemned urban planning for increasingly serving as a mechanism of control to perpetuate power rather than prompt social change and enhance living quality (Angotti, 2018; Simmie, 1974). For example, Jane Jacobs (1961) pioneeringly denounced the 1950s urban reconstructions in New York for displacing communities and creating isolated neighbourhoods. Urban planners and architects advocated for re-establishing the 'sense of place' to harbour the human need for belonging, safety and orientation. Placemaking gained a reputation as a solution that honoured the city's diverse and intricate fabric by reimagining public spaces and reinforcing connections between places and communities.

In the 1970s, urban designers and social critics continued to champion placemaking. William H. Whyte (1980) promoted bottom-up place design through meticulous and longitudinal fieldwork, such as observation, famously employing time-lapse photography to study public behaviour and space usage. Detailed observation informed evidence-based design principles, such as providing shading and movable seatings. The establishment of the first placemaking agency, the Project for Public Spaces (PPS), in the US during the

1970s, primarily grounded in Whyte's methods and findings, has since advocated for community-oriented, grassroots approaches to place-related issues and decision-making.

Since the 1990s, state and local governments have engaged in placemaking industries through creative placemaking, which leverages arts, cultural and creative thinking to enhance a place's appeal. Local governments recognised that infrastructure provision and design excellence alone could not cultivate a sustainable sense of place. Consequently, the placemaking agenda shifted towards place engagement through arts and cultural events, a practice known as creative placemaking (Markusen & Gadwa, 2010). Creative placemaking offered pathways for public artists, curators, designers and cultural programmers to collaborate as placemakers, generating projects with communities for governmental and corporate clientele (Griffiths & Barbour, 2016). The rise of the 'creative class' (Florida, 2002) encouraged governments to explore their governance philosophies by activating urban spaces. Public officials turned to creative clusters, precincts and cities to synergise geographically connected arts and cultural institutions. Government-funded arts and cultural initiatives began to restructure the urban economy profoundly (Flew, 2011), contributing to a transition from a manufacturing-based economy to an information-based economy that values knowledge and innovation (Peters et al., 2009). For example, Richards and Duif (2019) explored the transformative potential of creative placemaking in small cities, highlighting how cultural programming and strategic branding can stimulate economic, cultural and social development with an in-depth case of 's-Hertogenbosch, a small Dutch city and the birthplace of 16th-century painter Hieronymus Bosch.

2.1.2 *Placemaking in Australia, China and Taiwan*

On a global scale, placemaking does not present uniformity and varies in regional nomenclatures and terminologies. In Australia, placemaking has evolved into a relatively established field. Beginning in the early 1980s, design professionals and community groups collaborated to rejuvenate underutilised public spaces. Placemaking firms and agencies emerged to help activate Melbourne's public spaces, such as the laneways, through food, fashion and festivals (Hes & Hernandez-Santin, 2020). These agencies educated key stakeholders—academics, practitioners, community partners, interest groups and local governments—on enhancing place liveability and community engagement. A notable collaboration between the City of Melbourne (COM) and Danish urbanist Jan Gehl in the 1990s aimed to revitalise the cityscape, plagued by modernist planning and automobile dominance. The resulting report, *Places for People: Melbourne City 1994* (City of Melbourne & Gehl, 1994), established a benchmark for future policies with a community-centric perspective. Since the 2000s, placemaking has gained recognition as a public-facing issue in most Australian states and territories, with public consultation mandated for land-use decisions.[2] The Victorian Government released an engagement framework in 2022 to institutionalise public participation in local governments' placemaking strategies. Many local governments have adopted the International Association for Public Participation (IAP2) guidelines, inspired by Arnstein's (1969) model, to integrate public participation into daily administration.

In China, placemaking has undergone a compressed, transcultural process. Public spaces and facilities, such as parks, museums, libraries and squares, began appearing towards the end of the Qing Dynasty[3] and Republican China in foreign Concessions of major port cities, where scientific planning using investigation and survey tools was advocated by urban officials (Friedmann, 2007). Following the long hiatus of the multiple wars and the de-urbanisation period under Mao, the notion of public space was introduced to Chinese urban planners and designers in the 1980s, alongside the works of Kevin Lynch, Jane Jacobs and Christopher Alexander, who prioritised human experience in social places. Geographer Piper Gaubatz (2008) observed that new Chinese urban public spaces have become increasingly commercialised, with everyday use of open spaces, parks, and venues mingling with ephemeral grassroots spatial production and practices. Shopping malls, live houses and urban festivals are increasingly popular destinations in the burgeoning urban economy. In 2018, the City of Wuhan hosted the International Placemaking Week, resulting from a three-year partnership between UN-Habitat and public institutions like PPS, to explore community engagement in placemaking to improve societal health, happiness and well-being. The event launched the China Placemaking Network and witnessed the signing of the Wuhan Declaration to promote placemaking and quality public spaces in China.

In Taiwan, placemaking originated from the Japanese town-making movement (*machizukuri*), which aimed to restore urban and rural historical landscapes and nostalgic neighbourhoods through citizen participation in the town-making councils. In 1994, building on the Japanese experience, Taiwan launched the 'Integrated Community Building Program' by Chen Chi-nan, the vice-chairman of the Council of Cultural Affairs and an anthropologist. In contrast to China's 'state-sponsored market-driven' approach, Taiwan's placemaking underpinned 'government-led local-centred cultural development' celebrating Taiwan's unique cultural heritage and identity (Chan, 2011). This approach amalgamated five fields—people, culture, place, scenery and industry—to promote community autonomy at the core of community building (Zeng, 2007). Culture-led urban regeneration has been recognised as a strategy to enhance global competitiveness and improve economic performance, giving rise to urban institutions like the Taiwan Overall Community Making Association and Urban Regeneration Station (Chiu et al., 2019). Creative placemaking has emerged to harness the socio-cultural competitiveness of 'not so global' Taiwanese cities and address the crises of capitalist urban transformation. The Taipei City Regeneration Office engaged world-renowned consultant Charles Landry to enhance its cultural appeal through various agendas and events (Lin & Chiu, 2019). Another example,

Treasury Hill, has 'musemified' (Rogelja, 2021) from an illegal settlement to a curated and publicly administered artist village, showcasing Taipei's authenticity and creativity until today.

2.1.3 Existing Trajectories of Placemaking

Regardless of regional variations, three trajectories merit elaboration from the above account. First, contemporary placemaking practices disrupt the modernist nexus between settlement form and life experience by emphasising community over construction. This entails balancing pedestrian needs and place liveability with automobiles, skyscrapers and expansive transportation systems while cultivating a sense of community and ensuring its longevity through diverse placemaking techniques. This emergent community-centric theme has often bifurcated placemaking into two oppositions. 'Formal placemaking' projects are perceived to be spearheaded by institutional and expert systems, including the government, developers and planners. Characterised by the imposition of preconceived master plans for urban regeneration and redevelopment, formal placemaking leverages zoning laws and other regulatory mechanisms (Sweeney et al., 2018), encompassing state-led regeneration plans, urban precincts, redeveloped public spaces, public art and street furniture. Conversely, 'informal placemaking' initiatives engender what Lefebvre (1991) termed 'lived spaces'—organic and spontaneous initiatives regardless of institutional validation. Do-it-yourself (DIY) and tactical urbanism, diasporic navigation, protest constructions and protest art represent subjective placemaking endeavours, or 'retrospective world building' (Basso, 1996, p. 5). Community action addresses the power divide between placemaking as expert systems and placemaking as inhabitation, experience and interpretation (Lefebvre, 1996), offering a corrective framework through tactical urbanism (Finn, 2014). Recognising ordinary citizens as 'informal' placemakers revives a progressive notion of place as socially co-produced and acknowledges citizen agency and ownership of urban spaces.

Second, public participation gains significance as a central theme in placemaking processes, usually staged in opposing themes of democracy versus technocracy. Urban planners advocated for participatory planning to incorporate diverse resident perspectives (Hague & Jenkins, 2005). This orientation builds on a historical continuum of initiatives aimed at public inclusion in urban planning and design, such as community development, people-centric design and community modelling. Although implementation varies and exceptions occur, participation is increasingly seen as a hallmark of 'good governance'.[4] As governments have shifted towards a service-oriented approach since the 1990s, decision-making gradually embraced deliberative processes. Even in non-democracies, public participation in urban decision-making has been incrementally adopted through town hall sessions or public hearings by placemaking organisations. Non-electoral participatory mechanisms have been employed for public constructions in progressive Chinese provinces like

Zhejiang since the 1990s to enhance governance efficacy (Pavlićević, 2019). A series of laws and governmental guidelines in China encourage embedding the community voices into urban design, construction, governance and evaluation processes, including the Urban and Rural Planning Law (2007), Community Service System Building Plan (2011–2015) and Holistic Living Community Building Guide (2021, p. 53[5]). If the uneven distribution of citizenship rights has been conditioned by a city's geographical and demographic exclusion (Lefebvre, 1970/2003), placemaking perpetually renegotiates the political boundaries of urban space by adjusting and enacting inclusion.

Lastly, placemaking transcends discrete projects imposing material changes on the landscape, constituting ongoing processes animated by diverse actors, actions and aspirations on contested grounds. Current discourse often views placemaking through the lens of design industries, emphasising grand visions that culminate in place creation. Yet, in reality, placemaking unfolds through prolonged processes and conflicts: how are places maintained, activated, and rejuvenated? How do they navigate public contention? How are decisions about places made amidst differences? Placemaking is also increasingly seen in the aspiring utterance of place diversity, equity, inclusion and sustainability, and is associated with the increasingly important questions: Who gets to decide what becomes of the place next? What constituents are relevant? Who has been excluded and how is such exclusion rectified? Despite the substantial body of research produced by urban planning and design, placemaking should not be solely infrastructure-oriented or dominated by a single industrial sector. Matthew Carmona (2014) proposed a placemaking continuum, encompassing the 'knowing' phases of design and development phases and the 'unknowing' use and management phases.

Contrary to professionals' beliefs, urban planning, designing, development, construction, creative industry, cultural programming, place management, governance and inhabitation all contribute to placemaking and effecting changes. Balancing these elements is key to mediating issues between master planning and place-specific design, public good and private interest in partnerships, heritage protection and future development, multicultural and national interest, cosmopolitanism and indigeneity, democracy and technocracy. Not merely an instrument of harmony, placemaking can become a vector for contention if any of these issues is debated. For example, some scholars may prioritise preserving a place's cultural grounding and historical identity, whereas others may criticise conservative tendencies and warn against underestimating the plurality of competing visions for a place (Palermo & Ponzini, 2014).

2.2 Digital Placemaking: Interface of Urban Politics

2.2.1 Contextualising Digital Placemaking (1990s Onwards)

Since the 1990s, the placemaking industries have embraced the digital era through the widespread adoption of information and communication technology (ICT) in governmental, commercial and domestic settings. Digital placemaking enables to 're-plac[e] the city' through mundane and mediated practices, rendering the city unique, knowable and desirable (Halegoua, 2020, p. 5).

New actors have gravitated toward the nascent field of digital placemaking: urban consultancies, marketing firms, urban informatics, media architecture, social media platforms and tech companies. Urban practitioners and consulting firms were the pioneers in adopting the term 'digital placemaking', and consulting agencies swiftly emerged in the UK, the US and Australia.[6] These urban consultancies typically conceptualised digital placemaking as a suite of public-facing business strategies. They envision it as a tool to create attractive locations and fortify communities for both private and public initiatives. Media technologies, such as urban digital screens, location-based social networks and art installations, facilitate digital placemaking. However, in practice, digital placemaking was often superficially regarded as migrating placemaking information online and onto public screens.

The term 'digital placemaking' became a buzzword in 2017 with the Media Architecture Biennale's publication *Digital Placemaking Compendium* (Hespanhol et al., 2017). Media architecture focuses on the physical display, installation or projection of multimedia technologies in public spaces (Haeusler et al., 2012). Martin Tomitsch (2016) identified three orientations of digital placemaking: spectacle, infrastructure and community placemaking. Spectacle placemaking involves creating temporary urban destinations with light festivals, media art installations and seasonal events, which can boost tourism and the economy by transforming the urban public space into a stage and fostering 'ubiquitous interaction' (Hespanhol & Tomitsch, 2014). Infrastructure placemaking integrates ICTs into the urban architectural fabric, enabling the public to interact with interactive infrastructures through clicking, swiping and conversing, thereby improving the quality and efficacy of public services. Community placemaking, originating from the art scholarship regarding grassroots disobedience, digital activism and spatial aesthetic infrastructures (e.g., Bourriaud, 2002; Papastergiadias, 2006), facilitates collaborative decision-making by engaging stakeholders and local communities.

Smart mobile media and social networks have significantly transformed informal digital placemaking practices (Wilken & Goggin, 2013). For example, mobile devices and meet-up sites were used by expatriates to establish a cosmopolitan sense of belonging rather than reconnecting with or reconstructing home from abroad (Polson, 2015). Digital platforms like Weibo have played a role in constructing citizen identity and enabling digital placemaking as a communicative process that reinforces grassroots textual and symbolic reterritorialisation against the state's attempt to deterritorialise local identity and sense of place during China's rapid process of urban redevelopment (Wang, 2019, p. 43). Similarly, Germaine Halegoua (2020) emphasised that digital technologies facilitate citizens in habitually re-embedding themselves and recreating a sense of place imbued with unique meanings and deep affection, rather than fostering separation and alienation. The new generation of short video platforms like TikTok and Red has stimulated urban economic growth in *wanghong* (internet celebrity) cities by encouraging physical and virtual check-ins through digital storytelling and materialising users' emotional value into the agenda of wanghong urbanism (Cao, 2024).

Urban human–computer interaction (HCI) addresses urban interface designs, participatory technologies and big data modelling through task-based informatics. Urban interfaces can range from a small gaming app to a comprehensive smart city data hub. For instance, location-based games have been found to increase community awareness and assist community exploration and placemaking via gamified wayfinding and knowledge-building (Pang et al., 2020). Smart cities serve as experimental sites for urban decision-making through hybrid urban interfaces, including digital practices (e.g., selfies), data-gathering strategies (e.g., surveillance cameras) and data analytics. Other projects have investigated big data modelling and visualisation for evaluation and solutions to sustainable urban futures (Foth et al., 2015; Kitchin et al., 2017; Offenhuber & Ratti, 2014). Urban artificial intelligence (AI) and platform-propelled technologies and interfaces such as City Brain, digital twins and place-scoring systems are rapidly transforming the everyday management and urban operations (Barns, 2020; Cugurullo et al., 2024).

Less visibly, tech companies and global brands have also entered the scene through urban digital transformation and redevelopment. From sensorially stimulating flagship stores to smart precincts, there is a global trend for tech companies to propose urban innovation solutions and transform urban public spaces amidst popular pushbacks. A high-profile example is the Quayside project by Sidewalk Lab, Google's urban planning and infrastructure subsidiary. Sidewalk Labs proposed building a smart precinct in Toronto and managing service delivery in lieu of the government (Hodson & McMeekin, 2021). It was abandoned in 2020 amidst citizen protests against urban surveillance and data abuse, citing 'financial uncertainty during the pandemic'. Other instances include public rejection of the Google campus in Berlin, citizens' pushback on Amazon's proposed headquarters in New York City, public dispute around Apple Federation Square in Melbourne and debates over tech moguls creating new cities in the United Arab Emirates and San Francisco. In a *New York Times* guest essay, urbanists Edward Glaeser and Carlo Ratti (2023) rightly argued that cities built by investors and corporations are not morally inferior to those planned by the governments and may even push the boundaries of sustainable building solutions. However, akin to social networks and platforms often lauded as virtual 'town squares', tech-driven cities raise concerns about privacy, security and accessibility due to their potential to create surveillance architecture and use harvested data to predict public behaviour, apply penalties, twist public debates and exploit decision-making (Flynn & Mackay, 2019). These multidimensional 'turf wars' between global tech corporations and urban public spaces remain underexplored and have not yet sparked sufficient debates like their virtual counterparts.

2.2.2 *Governance Frameworks of Digital Placemaking in Australia, China and Taiwan*

Digital placemaking in Australia has coincided with the institutionalisation of the engagement sector, increasingly integrating data tools into urban governance. In 2018, the COM Council announced a new full-time position, 'digital placemaking lead', within its placemaking and engagement branch. This role, framed through the lens of 'place,' would spearhead the online engagement portal, Participate Melbourne (https://participate.melbourne.vic.gov.au/), to foster internal project integration and enhance external stakeholder relations. Participate Melbourne is one node in an expanding matrix of state and local government-owned engagement portals, including Engage Victoria (https://engage.vic.gov.au/), Connect Stonnington (https://www.connectstonnington.vic.gov.au/), Conversations Moreland (https://conversations.moreland.vic.gov.au/) and many others. These portals originated from the Victorian Government's Public Engagement and Partnership Vision (2018), aiming to elevate community and system outcomes across 79 Victorian local governments. While the establishment of digital public infrastructure alone does not provide a panacea for existing urban challenges, it constitutes a pivotal step in renegotiating power dynamics between citizens and government, thereby restoring political trust and reinforcing the values of urban governance. On a national and strategic level, government agency Infrastructure Australia has adopted a 'digital by default' policy, mandating that all federally funded infrastructure projects incorporate a digital twin and digital asset management component within the next decade as part of its 2021 Australian Infrastructure Plan.

Over the past decade, China has led the charge in urban digital transformation by radically platformising urban governance and leveraging big data and AI. In 2020, Shanghai unveiled a municipal platform, Single Platform for Urban Management (*yiwangtongguan*), for centralised service provision in areas such as public security, market supervision, environmental protection, emergency response and daily urban management (Todaro, 2023, p. 95). Terabytes of real-time data are harvested from 'surveillance cameras, traffic monitoring systems, and a network of approximately 5.1 million Internet of Things sensors that collect more than 31 million data daily' by 'more than 70 government departments at the municipal and district levels; and one municipal-level, 16 district-level and 215 subdistrict-level urban operation management centres' (Todaro, 2023, p. 95). The Shanghai model has been swiftly adopted by other cities, such as Beijing and Guangzhou (Xiong, 2022). The vast volume, diversity and granularity of data, enabled by the highly integrated digital architecture designed and controlled by the government, facilitate coordinated urban management. Augmented by distributed personal smart devices, this data can be instantly operationalised to track and map urban mobility, as evidenced by the rollout of 'health codes' during the Covid-19 pandemic (UN-Habitat China, 2020).

Taiwan has garnered acclaim as a beacon for digital democracy, operating on parallel systems supported by public administrators and civic hackers. On the government side, programmer and activist Audrey Tang became the first Digital Minister without Portfolio in the Legislative Yuan in 2016 and the inaugural Minister for the Ministry of Digital Affairs in August 2022. The government launched the Regulatory Sandbox platform in 2017, inviting tech entrepreneurs and social innovators to experiment and drive regulatory changes. The Taiwan government manages the national participation platform, Join (https://join.gov.tw/), for public matters. Four million participants out of Taiwan's 23 million population have registered on the platform, and anyone can propose an agenda for public debate. For any e-petition garnering more than 5,000 signatures, the relevant ministries must publicly respond (Horton, 2018). Civic hackers formed a politically minded civic tech community, g0v (gov zero, https://g0v.tw/), playing active roles in the 2014 Sunflower Movement and building a citizen auditing system for Taiwan's central government budget from publicly available data. The g0v community initiated the vTaiwan platform (https://vtaiwan.tw/) to promote government transparency and public participation in the legislative process. Using an AI system to manage the conversation, vTaiwan has successfully facilitated rough consensus among multiple stakeholders towards implementing the 'Diversified Taxi Services Plan' in 2016 (Robbins & Tang, 2022). Since 2018, the annual presidential civic hackathon has encouraged civic tech communities to use open data, technology and creative thinking to solve societal problems and promote collaboration. By integrating technology into citizens' lived spaces, g0v has provided proof of concept for open-source, open government initiatives in Hong Kong and Italy.

2.2.3 Gaps in Existing Digital Placemaking Studies

Surveying the transdisciplinary expanse of digital placemaking, I conceptualise it as the dynamic integration and convergence of digital media technologies into both established and emergent placemaking industries. This concept remains fluid and boundless, continually attracting new actors and innovative practices. These entrants, in turn, refine the definition of digital placemaking by broadening its practices, industries and scales, demonstrating the necessary fuzziness (Kitchin, 2022) between conceptualisation and actualisation.

Consequently, digital placemaking intersects with the smart city discourse, sharing the ambition of integrating technology-driven solutions into the urban fabric to achieve common goals such as enhanced quality, efficiency and sustainability. However, their scope and focus diverge. Smart city initiatives, with their infrastructure-centric and development-oriented focus, encompass the full spectrum of urban systems—transportation, security, economy, environment and data management (Willis & Aurigi, 2017; Zhao et al., 2021). In contrast, digital placemaking prioritises forging connections and cultivating meaning between people and places across expansive, scalable contexts. The adaptability and community-driven ethos render digital placemaking more affordable and less reliant on large proprietary technical systems and high-level authorisation often implied in the 'actually existing' smart cities (Shelton, Zook & Wiig, 2015). Digital placemaking resonates more deeply with the concept of placeness, evoking a richer tapestry of cultural, social, historical and affective connections, while smart cities are often perceived as products of political economies and technical regimes (Kitchin, 2023). The notion of digital placemaking necessitates deeper exploration into the depth and sophistication of digital democracy and the human condition.

Existing digital placemaking literature has limitations, hindering its evolution into a mature research field. At present, digital placemaking research relies heavily on empirical descriptions to document emerging practices and bourgeoning theoretical groundings. To date, two book-length publications on the topic (Halegoua & Polson, 2021; Hespanhol et al., 2017) function as casebooks, examining digital placemaking through specific technological applications and evaluated media technologies for particular ends. These bodies of literature illustrate digital placemaking primarily through lived experiences, such as media façade, art installations, travel selfies, livestreaming, hybrid street furniture, social media, data management and mobile media apps and digital practices. The narrow focus creates a fragmented and ad hoc academic reality that contradicts the comprehensive scope of place and placemaking literature. Digital placemaking studies must move on from the research logic that treats individual facets as representative of the entire phenomenon. Instead, they should acknowledge urban complexity, where diverse digital placemaking practices co-exist and reflect heterogeneous urban agendas and political aspirations. Digital placemaking research must find an intellectual home that genuinely honours its transdisciplinary, convergent and emergent qualities.

Further, digital placemaking research must overcome the naivety reminiscent of the 'democratising thesis' in early digital media studies of the 1990s. Numerous cases highlighted the governmental adoption of digital placemaking as a 'paradigm shift' from linear, single-actor, top-down spatial governance to discursive, multiple-actor, decentralised stakeholder engagement (e.g., Toland et al., 2020). Simultaneously, digital placemaking has been narrowly celebrated as 'bottom-up', grassroots actions, falsely contrasting the 'top-down', neoliberal smart cities. Prior to the special *Convergence* issue, digital placemaking was rarely associated with developer manipulation and government control (e.g., Özkul, 2021; Witteborn, 2021), which increasingly represent the real existing digital placemaking landscape. For digital placemaking research to advance, it requires the theoretical sophistication found in media theories and science and technology studies. I contend that digital placemaking studies must overcome current methodological limitations and avoid producing dichotomised, disconnected and exaggerated perspectives.

2.2.4 Digital Placemaking as Political Interface of Digital Cities

The political implications of digital placemaking remain insufficiently examined. As a crucial interface within contemporary digital cities, digital placemaking demands critical reflection. While placemaking studies have evolved to challenge dichotomous categorisation (e.g., Courage et al., 2020), digital placemaking remains ensnared in binary perceptions. For example, Marcus Foth (2017) delineates digital placemaking into 'accelerated' and 'natural' categories. The former arises from the ambitions of developers, construction companies and city officials to rapidly invigorate new urban developments and expedite the creation of vibrant communities. In contrast, the latter refers to the organic lifecycle of buildings as they become 'occupied, inhabited, decorated, customised and renovated' (Foth, 2017, p. 203). Although Foth correctly identifies the varying 'speed-scapes' of placemaking, his dichotomy neglects the eclectic role of economic actors, such as place managers and public organisations, who defy simplistic categorisation as 'natural' or 'accelerated'. These actors extend the temporal dimension of longevity by preserving places through an economy of care and maintenance. Overlooking this perspective underestimates the relentless labour essential to sustaining the materiality of digital placemaking.

Digital placemaking introduces a novel arena where public participation becomes enmeshed in government programs via policy planning, platform mediation and technological intervention. It increasingly serves as an 'invited space' (Miraftab, 2004) created by institutions to mediate and resolve conflicts among diverse urban stakeholders, as opposed to an 'invented space' where grassroots organisations unilaterally challenge the status quo. This shift illustrates the everyday ramifications of public participation beyond electoral cycles, overthrowing previous theoretical frameworks that view public power as dormant in non-election times (e.g., Sartori, 1987). The ubiquity of digital platforms and networks has rendered what was once perceived as tactical placemaking more observable and impactful in digital cities (Halegoua, 2020). McQuire (2017) posits that the intersection of the urban and the digital constitutes a pivotal site for a new order, linking civic and cultural practices to the 'industrial' logic of digital technologies. Digital placemaking must navigate a more intricate and contested milieu, as contemporary urban space production attracts a wide array of civic, commercial and governmental interests.

Thus, the power dynamics that permeate digital placemaking are splintered (Graham & Marvin, 2002), kaleidoscopic and disorienting. New forms of conflicts manifest through ubiquitous media infrastructure is portable, real-time and geotagged, materialising in hybrid forms of hashtag activism, DIY urbanism and various citizen engagement activities (Sweeney et al., 2018). The discourse is increasingly place-based and contextually responsive. Digital placemaking as a process of 'commoning' multiple spaces, temporalities and relations where technocapitalist logics transcend physical geography demonstrates that the power dynamic in digital cities is neither binary nor linear. New conceptual models are imperative for digital placemaking practices to capture the new interactions across a broad spectrum of placemakers while honouring the multitude of digital urbanism. As the former President of Taiwan, Tsai Ing-wen, reflected at her 2016 inauguration, 'Before, democracy was a showdown between two opposing values. Now, democracy is a conversation between many diverse values' (Tang, 2019). There is an urgent need to explore the intersections and complex processes where diverse initiatives collide and converge to form an unstable nexus.

2.3 Tripartite Assemblage-Based Model

2.3.1 Urban Assemblage, Place and Placemaking

Urban assemblage thinking, originating from Deleuze and Guattari (1987), employed the concept of *agencement* (assemblage) to illuminate the interplay of human and non-human components within a rhizomatic network, contrasting it with the hierarchical tree-like knowledge system. The former establishes multiple, non-hierarchical connections among heterogeneous entities, while the latter operates with rigid hierarchies and binaries. For planner and theorist Kim Dovey (2012), an urban assemblage represents the confluence of interconnectivity and flows among constituent parts. Diverse components—such as urban life, traffic, goods, money, plans, policies, nature and infrastructure—amalgamate to give the street intensity and an emergent sense of place. These components are perpetually engaged in processes of 'coming together (territorialisation)' and 'falling apart (deterritorialisation)' (Deleuze & Guattari, 1987, p. 406). Urban assemblage elucidates the contingent processes through which heterogeneous flows perform, cohere, sustain, dissipate and disintegrate, thereby reflecting the complexity, contingency and labour intrinsic to placemaking processes that actualise socio-material transformations (Bakawa Country et al., 2016).

Urban assemblage thinking respects the heterogeneous and multitudinous time-space conditioning of digital placemaking. It refrains from reducing digital placemaking to a mere operational task with defined start and end dates in a specific geographical area. Instead, it renews an essentialist and territorial place ontology with openness and fluidity. Thus, digital placemaking can occur at various scales—building, street, neighbourhood, precinct, city, nation-state and beyond—traceable through the amalgamation and disintegration of heterogeneous actors and actions within a place's life cycle. Each assemblage uncovers distinct associations, relations and issues. No assemblage supersedes the others, just as no single industry dominates the entire placemaking process. Urban planning, design, development, governance and inhabitation are embedded within the synergistic morphology of digital placemaking assemblages.

I employ 'assemblage' as an active verb to prioritise juxtaposing heterogeneous elements traditionally perceived in isolation. Place and placemaking are inseparable, relational processes. Places are not mere physical sites, being produced and sustained through actions. Human geographer Jon Anderson (2012) asserts that a surfed wave exists through the practices and coordinated efforts of the surfer's body, ocean waves, board and cord, weather, surfer's luck and unpredictable currents, much like the coastline coheres and stabilises as a place over time, through many ocean waves, many surfers and their surfing actions. Urban assemblage thinking considers 'the material, actual and assembled' and 'the emergent, processual and multiple' (Farías, 2009, p. 15). Place and placemaking intertwine through repeated practices and incessant labour (Sweeney et al., 2018). Tensions emerge when conflicting relationships manifest during these processes, emphasising how various urban segments, departments and institutions align and collide over time, enabling certain urban peculiarities to prevail over others (McFarlane, 2011). The mercurial nature of urban assemblages allows for the consideration of urban nuance, relations and fluidity.

Second, I utilise urban assemblage to inquire into, rather than critique, the political implications of digital placemaking. As assemblage thinking prioritises inquiry in the conditions of creative production over knowledge *qua* representation (Rajchman, 1998), it facilitates an analytical approach that neither presumes nor disregards dominance. I initiated the project with no predefined objects of criticism in mind, instead treating all actants in the assemblage as subjects of wonder and surprise. Embracing unexpected connections and unconventional perspectives, the assemblage lens rejects assumptions of universal laws and predictable outcomes (McFarlane, 2011). This perspective precludes the dismissal of tactical urbanism as 'hyperlocal' initiatives incapable of offering genuine alternatives to neoliberal urbanism, as critical urbanists might argue (Brenner, 2020). Instead, I demonstrate that the 'large scale', 'worldwide' nature and 'broader implications' of the global hierarchy do not diminish the vibrancy and significance of the 'local', 'informal' and 'microspatialities' (Dovey, 2011, p. 348). Urban assemblage allows me to construct what I term the ouroboros of digital placemaking, an oblique analytical tool that circumvents binary and structural presumptions of urban politics.

2.3.2 *The Ouroboros of Digital Placemaking*

When Bruno Latour (2007, 2013) conceptualised the 'political circle' to define politics as the process of an issue 'becoming political' rather than the content of political debates,[7] he simultaneously rebelled against and saluted his predecessors. For Latour, issues precede politics. Issues are debatable matters of concern rather than risk-free facts that end discussions. The 'matters of concern are not firmly substantiated, prone to unintended consequences and deeply intertwined with particular situations' (Chhachhi & Donovan, 2014, p. 881). Arising from the specificities of reality and happenings, issues agitate, complicate and provoke speech. Politics is thus a secondary indicator of the stage of a specific issue in the political circle. Latour described issue-centred politics as *Dingpolitik*, which revolves around issues that attract specific public cohorts to sustain the debate. This conception inverts the conventional human-centred *Realpolitik*, where issues automatically enter a pre-populated political arena and await solutions or judgements. For Latour, politics commences when new human and non-human associations confront society with unexpected controversies (political-1). Politics persists when issues disturb the public and mobilise the formation of various assemblies (political-2). Politics become institutionalised when governments recognise and allocate the problem to established government institutions and agencies (political-3) followed by public deliberations (political-4). Politics dissipates with a government's incorporation of the issue (or failure to do so) as part of the daily administration (political-5).

Notably, the political circle allows an issue to migrate from latent political realms (as emergence in political-1 and quotidian administration in political-5) to overtly political realms (as public controversies in political-2, sovereign issues in political-3 and deliberations in political-4) and back. The fivefold lifespan of political issues integrates different forms of public participation, juxtaposing the pragmatist public participation in the tradition of Walter Lippmann and John Dewey (political-2) and Habermasian deliberative public participation (political-4). The political circle does not depict a hierarchical echelon of power but accounts for the trajectory of an issue by showing how heterogeneous actors from complex systems shape a concern. Particularly, political-1 and political-5 render visibility to an integral part of politics conventionally deemed apolitical and routinised public administration. In this way, it extends the politics beyond the high-profile moments of public discontent into a trajectory where the longevity of the public concern becomes traceable and solutions evaluable.

Modifying Latour's (2007) political circle, I propose to prototype the politics of digital placemaking as an ouroboros, an iterative process that potentiates multimodal political reflexivity. First, I exchange the circle with the ouroboros, a snake biting its own tail, an ancient symbol for cyclic renewal. The ouroboros connects political-1 and political-5 in line with Latour's original conception. It does not imply a single-direction transformation. Rather, it can move clockwise or counterclockwise or leap from one stage to another under certain circumstances. The ouroboros illustrates an iterative trajectory of the procession of events within the potential of reflexivity, rebirth or demise. Second, I integrate into Latour's political stages three interweaving assemblages: activist, organisational and governmental. Politics in digital cities emerges as new human/non-human associations are formed in political-1. Public discontents are expressed and performed in various ways in the activist assemblage in political-2. The governmental assemblage acknowledges and transforms the contested issue into a public-facing question to be addressed through established institutions and policy frameworks in political-3. All assemblages are mobilised through various forms of public consultation and deliberation to deliver solutions in political-4. The organisational assemblage incorporates the solution to the issue and routinises it as daily administration in political-5. Each assemblage acts as a Latourian *oligopticon*, vividly revealing how the micro-structures assemble macro-phenomena.

At its most reductive interpretation, an *activist assemblage* comprises the countercurrents publicly performed and persistently embodied through networked corporeality. It can manifest as Michael Warner's (2002) 'counterpublics', or what Manuel Castells (2012) termed a 'network of outrage and hope' that challenges the status quo. This assemblage encapsulates traits of urban protests and networked activism that underpin interpersonal connections and build on frames of intimacy. The activist assemblage is a ceaseless entanglement of placards, hashtags, social media campaigns, alternative art events, media, rhetoric, personal enthusiasm, collective belief, cultural production, petitions, calculations, tactics, freedom of speech, censorship, public concerns, disobedience, narratives and counter-narratives, performativity, affect, commentators, public intellectuals, colleague professors, artists, musicians, local politicians, sometimes Members of Parliament, networks of activists, campaign volunteers, content contributors, advocacy organisations and non-profits. These activist assemblages are hardly mere grassroots initiatives, as actants may range from unpaid armature activists to professional advocacy groups. Not all activist assemblages revolve around righteous or legitimate causes, nor do all advance to the next political stage. By blurring formal and informal placemaking processes, activist assemblages offer ordinary people avenues for participating in space, claiming rights and disrupting political establishment.

An *organisational assemblage* is tasked with sustaining place vibrance and fostering economic prospects through the onerous labour of place management. Hannah Arendt (1958) defines labour as a cyclical, repeated process imbued with necessity. An organisational assemblage is necessary to prevent deterioration, cover running costs, maintain operations and support potential growth. The organisational assemblage inevitably encompasses but is unlimited to business cases, implementation of business plans, vision and mission statements, Key Performance Indicators, evaluation matrices, annual reports and statements, legal agreement, cash flows, markets, profit margins, infrastructural amortisation and depreciation, stakeholder analysis and management, strategic partners, digital marketing, clientele avatars, influencers, urban marketing agencies, heat maps, censors, consulting firms that commercialise business solutions during urban digital transformation, social media platforms, cross-platform strategies, algorithms, AI, big data modelling, tech companies, city councils, local businesses, public relation, travel agencies, tourists, the C-suite, cleaners and place managers. While incorporating market-driven actants and actions, an organisational assemblage cannot be mechanically reduced to a neoliberal frame. Ash Amin and Nigel Thrift (2007) argue that business decisions in new urban economies intertwined with globalisation, digital technologies and networked relationships are not driven solely by rational, profit-maximising actors. Cultural, emotional and embodied aspects, such as passion, moral values, soft skills and trust, also shape economic behaviour. The vantage point of organisational assemblage resists reducing city dwellers to mere consumers in the routines of capitalised urban circuits and occasional voters in neighbourhood corporations.

The *governmental assemblage* recognises public turbulence as an urban public-faced issue that requires allocation to specific ministerial portfolios, departments and policy fields. Structurally integrating the issue (rather than the solution) into systems of checks and balances and due processes is an act of institutionalisation. Samuel Huntington (1968) defines institutionalisation as a process by which organisations and procedures acquire value and stability. Emergent complex phenomena like digital placemaking that traverse diverse sectors do not readily fit into any policy catchment. The role of governmental assemblage is essential to recognising new problems and inventing new categories through constant policy and institutional innovation. To achieve this, the governmental assemblage must strategically manage urban aspirations, grounding them contextually in intersectional practices and local knowledge, with increasing affinity for broad engagement mechanisms. The governmental assemblage may encompass a mishmash of various levels of government, politicians, members of parliament, protocols, hierarchies, parliamentary processes, briefing and correspondences, policy papers, reviews, inquiries, reforms, stakeholder engagement events, budget bids and cycles, vision statements, business cases, evaluations, public accountability, consultation firms, communities of practices, interest groups and summits. Karl Mannheim (1937, p. 118) observed that 'the fundamental tendency of all bureaucratic thought is to turn all problems of politics into problems of administration'. While this observation highlights the limitations of bureaucratic compartmentalisation, it also underscores the administrative imperative of assigning issues to actionable trajectories and integrating solutions into business-as-usual processes, thus completing the ouroboros. In doing so, the governmental assemblage cannot be reduced to bureaucracy.

Rather than studying a bounded group or organisation, assemblage thinking enables the exploration of the interconnectedness and fluidity of group formations and their linkages to other groups not traditionally considered together. Activist, organisational and governmental assemblages reflect the views of—but are not limited to—the activists, place managers and governments. For instance, the activist assemblage diverges from an activist group. The former eschews strict and definitive boundaries, encompassing all activist-minded actants supporting the Latourian issue. Assemblage thinking situates the activist assemblage in relation to the non-activists, such as media professionals, general public, the organisational and governmental assemblages and the material, socio-technical urban environment that supports it. Similarly, the organisational assemblage differs from a business entity, and the governmental assemblage transcends partisan politics or bureaucracy. The assemblage lens describes more agile actors in the contemporary polity that reside in informal and flexible collaboration spaces rather than saturated within the formal institutions and political hierarchies. Instead of emphasising positions, the ethnography of the assemblage highlights relationships, where each actant can act as an intermediary between various political steps. The nomenclature informs my understanding of digital placemaking as an intricate mode of politics arising from the fluidity and synergies of different assemblages. The notion of politics as absolute opposition between irreconcilable enemies can be renewed by the clash of diverse concerns, connective styles and *modi operandi*.

The ouroboros elucidates digital placemaking's compositional heterogeneity, processual relationality and democratic potentiality. It reveals digital placemaking as trans-industrial practices and open-ended and multi-scalar processes. It facilitates the analysis of specific and evolving components of digital placemaking, including traditional 'things' such as urban structure and infrastructures, individual and collective human actants and emerging technological practices, trends and concerns that permeate the contemporary urban environments. Illustrating how these socio-technical assemblages interweave in contemporary urban milieu, the ouroboros interlocks assemblages organised around place use and activism, economy and governance. It underscores the significance of place managers, whose work in infrastructural maintenance, events programming and financial operations, though seemingly mundane, is indispensable to producing places through constant labour. It also echoes Matthew Carmona's (2014) 'place-shaping continuum' while foregrounding public discontent as a major factor in contemporary placemaking negotiations. It aids inquiries into new configurations rather than applying existing hierarchical and normative assumptions to the research.

The model also shows that different modes of public participation cut through placemaking assemblages and decision-making processes, affirming the ambient 'participatory condition' (Stalder, 2018). Public participation ceases to be a theoretical potential but rather an urban reality whose fruition cannot be reduced to the changing technological paradigm. Instead, it relies on numerous changes in placemaking, politics, culture and media. The multitude, necessity and problems of public participation warrant deeper reflection. The ouroboros captures different forms of urban participation, ranging from street performance, protest, consumption, online discussion, public consultation, petitioning and direct voting in urban affairs. This model does not unrealistically require an 'omnipotent' and omnipresent public set up by Lippmann (1922) as a straw figure to intervene in every issue whenever possible and to justify the need to manage public opinion. It also refrains from mystifying public participation as an absolute virtue. High-quality public participation necessitates devoted governments, responsive businesses and caring individuals who painstakingly navigate information overload and reach consensus on factual matters to effectively deliberate on 'matters of concern'. Lowered technological affordance and increased accessibility to participate in public affairs prove to be simultaneously a blessing and a curse. Different issues have diverse abilities in attracting public attention and generating various levels of support. The popular interest and 'buy-in' of a particular issue can only be maintained for different lengths of durations under different circumstances through various media platforms. Public participation constitutes the polity rather than dominates it, echoing political scientists' pragmatic observations that 'public participation at its best operates in synergy with representation and administration to yield more desirable practices and outcomes of collective decision making and action' (Fung, 2006, p. 26).

This remodelling exercise reimagines urban politics as multilateral repeated games, illuminating politics as ongoing transformations rather than temporary effects. In contrast, both Arnstein's (1969) ladder and Cardullo and Kitchin's (2019) scaffold depict politics as a single-shot game where the public confronts the institution, subjecting public participation to normative political opposition and hierarchical power structure. The ouroboros accounts for participation beyond momentary winning or losing without presuming the centrality of government or neoliberalism. Accounting for the stabilising power of the organisation assemblage, it honours labour and maintenance in digital placemaking and finds hope in its potential to further support an economy of care. This model enables investigating the circuitous politics of digital placemaking as surprises and experiments and escapes a binary imagination.

The ouroboros accounts for opportunities residing in latent modes of urban democracy by joining political-1 and political-5. The conventional understanding of politics typically incorporates only political-2 to political-4, rendering them insensitive to new yet ubiquitous practices like digital placemaking and smart cities. In contrast, the political circle can accurately reflect how some smart city technologies have been taken for granted as a highly sought-after method of public administration (political-5) without undergoing public discontent and discussions (political-4). The ouroboros provides a new vocabulary to trace the emergence of issues attributed to digital placemaking and their successive political stages. The ouroboros is sensitive to emerging technological trends and practices previously ignored and excluded from the political realms. It clarifies how apolitical issues become political and how overtly political issues might mutate over time and migrate into realms of management and administration.

Finally, my understanding of digital placemaking concerns more than one ouroboros but the intertwining of many ouroboroi, some of which are established while others are emerging. Since the ouroboros follows one issue, many ouroboroi can be 'stacked' to reflect urban complexities that may become disabling. This book will show that digital cities are messy bundles of legacy and emerging issues, ranging from finance, architecture, governance and technologies that play out in diverse socio-technical contexts, political settings and time-spaces. While the ouroboros demonstrates the trajectory of one matter of concern, I use the ouroboroi to emphasise digital placemaking as a heterogeneous and relational field where multiple matters of concern emerge, mature, resolve, stagnate, evolve or dissipate. The point of the analysis is to untangle the abstract complexities from concretised realities, in the hope of gaining consensus and fostering action.

Notes

1 Teahouse goers would 'tell ancient and modern stories, comment on society, play chess, gamble, criticise public figures, investigate private matters, and gossip about the secrets of the boudoir' (Wang, 2003, p. 95).
2 Australia has three levels of government, including federal, state and territory parliaments, and local councils. In most cases, state and territorial governments enact legislation on land-use planning frameworks and delegate implementation to local municipal authorities.
3 The first large-scale construction and expansion of 'space' was after the 1905 Preparative Constitutionalism when the Qing court sent five ministers to Europe to inspect Western politics. Their reports applauded public facilities such as museums, parks and libraries, and recommended that the capital, Beijing, should be an example of planning and managing similar public facilities to guide public life.
4 Public participation does not necessarily result in civil society, see Mattingly (2019). Similarly, democracy is not free of crisis—historically it has never been a failproof model for superior governance.
5 Holistic Living Community Building Guide has encouraged developing a community-led model as a 'communicative platform'. https://www.gov.cn/zhengce/zhengceku/2022-01/12/5667815/files/a84ca3d812e54074a43e332f3cc18eca.pdf.
6 Prominent examples include UK-based Calvium, Australia-based Digital Placemaking Institute and US-based PPS.
7 *Dingpolitik* is not merely about the institutional practices of parties, governments and parliaments as 'a way of codifying particular institutional and technical practices' (Barry, 2001, p. 201); it describes the 'ways in which artefacts, activities or practices become objects of contestation' (p. 201) or becomes political.

References

Amin, A., & Thrift, N. (2007). Cultural-economy and cities. *Progress in Human Geography*, *31*(2), 143–161.

Anderson, J. (2012). Relational places: The surfed wave as assemblage and convergence. *Environment and Planning D: Society and Space*, *30*(4), 570–587.

Angotti, T. (2018). *Metropolis 2000: Planning, poverty and politics* (Vol. 2). London and New York: Routledge.

Aravot, I. (2002). Back to phenomenological placemaking. *Journal of Urban Design*, 7(2), 201–212.

Arendt, H. (1958). *The human condition*. Chicago: University of Chicago Press.

Arnstein, S. R. (1969). A ladder of citizen participation. *Journal of the American Institute of Planners*, *35*(4), 216–224.

Augé, M. (1995). *Non-places: Introduction to an anthropology of supermodernity*. London and New York: Verso.

Barns, S. (2020). *Platform urbanism: Negotiating platform ecosystems in connected cities*. Singapore: Springer.

Barry, A. (2001). *Political machines: Governing a technological society*. London: A&C Black.
Basso, K. H. (1996). *Wisdom sits in places: Landscape and language among the Western Apache*. Albuquerque: University of New Mexico Press.
Bourriaud, N. (2002). *Relational aesthetics* (p. 44). (trans.) Pleasance, S, Woods, F, & Copeland, M. Dijon: Les presses du reel.
Brenner, N. (2020). Is "tactical urbanism" an alternative to neoliberal urbanism? In: C. Courage et al. (eds), *The Routledge handbooks of placemaking* (pp. 312–321). New York and London: Routledge, Taylor & Francis.
Cao, L. (2024). From online to onsite: *Wanghong* economy as the new engine driving China's urban development. *Environment and Planning A: Economy and Space*, *56*(4), 1061–1076.
Cardullo, P., & Kitchin, R. (2019). Being a 'citizen' in the smart city: Up and down the scaffold of smart citizen participation in Dublin, Ireland. *GeoJournal*, *84*(1), 1–13.
Carmona, M. (2014). The place-shaping continuum: A theory of urban design process. *Journal of Urban Design*, *19*(1), 2–36.
Castells, M. (2012). *Networks of outrage and hope. Social movements in the Internet age*. Cambridge, UK: Polity Press.
Chan, S. C. (2011). Cultural governance and place-making in Taiwan and China. *The China Quarterly*, *206*, 372–390.
Chhachhi, A. & Donovan, K. P. (2014). 'Development' as if we have never been modern: Fragments of a Latourian development studies. *Development and Change*, *45*(5), 869–894.
Chiu, Y. H., Lee, M. S., & Wang, J. W. (2019). Culture-led urban regeneration strategy: An evaluation of the management strategies and performance of urban regeneration stations in Taipei City. *Habitat International*, *86*, 1–9.
City of Melbourne, & Gehl, J. (1994). *Places for people*. Melbourne: City of Melbourne.
Courage, C., Borrup, T., Jackson, M. R., Legge, K., McKeown, A., Platt, L., & Schupbach, J. (Eds.). (2020). *The Routledge handbook of placemaking*. Routledge.
Country, Bakawa, Wright, S., Suchet-Pearson, S., Lloyd, K., Burarrwanga, L., Ganambarr, R., ... Sweeney, J. (2016). Co-becoming Bawaka: Towards a relational understanding of place/space. *Progress in Human Geography*, *40*(4), 455–475.
Cugurullo, F., Caprotti, F., Cook, M., Karvonen, A., McGuirk, P., & Marvin, S. (Eds.). (2024). *Artificial intelligence and the city: Urbanistic perspectives on AI*. London and New York: Taylor & Francis.
Deleuze, G., & Guattari, F. (1987). *A thousand plateaus: Capitalism and schizophrenia*. Bloomsbury Publishing. B. Massumi (tran). Minneapolis: University of Minnesota Press.
Dovey, K. (2011). Uprooting critical urbanism. *City*, *15*(3–4), 347–354.
Dovey, K. (2012). Informal urbanism and complex adaptive assemblage. *International Development Planning Review*, *34*(4), 349–368.
Farías, I. (2009). Introduction: decentring the object of urban studies. In I. Farias and T. Bender (eds), *Urban assemblages: How actor-network theory changes urban studies* (London: Routledge, 2009), pp. 1–24.
Finn, D. (2014). DIY urbanism: implications for cities. *Journal of Urbanism: International Research on Placemaking and Urban Sustainability*, 7(4), 381–398.
Flew, T. (2011). *The creative industries: Culture and policy*. Newcastle: Sage.
Florida, R. (2002). *The rise of the creative class* (Vol. 9). New York: Basic books.
Flynn, S., & Mackay, A. (2019). *Surveillance, architecture and control*. Berlin: Springer International Publishing.
Foth, M. (2017). Some thoughts on digital placemaking. In: Hespanhol, L., Haeusler, M.H., Tomitsch, M. & Tscherteu, G. (eds.), *Media architecture compendium: Digital placemaking*, 203–205. Stuttgart: av edition.

Foth, M., Brynskov, M., & Ojala, T. (2015). *Citizen's right to the digital city*. Berlin: Springer. 978–981.
Friedmann, J. (2007). Reflections on place and place-making in the cities of China. *International Journal of Urban and Regional Research*, *31*(2), 257–279.
Fung, A. (2006). Varieties of participation in complex governance. *Public Administration Review*, *66*, 66–75.
Gaubatz, P. (2008). New public space in urban China. Fewer walls, more malls in Beijing, Shanghai and Xining. *China Perspectives*, *2008*(2008/4), 72–83.
Gehl, J. (2013). *Cities for people*. Washington, DC: Island Press.
Glaeser, E & Ratti, C. (2023). Billionaire Built cities would be better than nothing. https://www.nytimes.com/2023/12/26/opinion/flannery-associates-billlionaires-woodlands.html?smid=nytcore-ios-share&referringSource=articleShare
Graham, S., & Marvin, S. (2002). *Splintering urbanism: Networked infrastructures, technological mobilities and the urban condition*. London and New York: Routledge.
Griffiths, M., & Barbour, K. (2016). *Making publics, making places*. Adelaide: University of Adelaide Press.
Haeusler, M. H., Tomitsch, M., Tscherteu, G., & Van Berkel, B. (2012). *New media facades: A global survey*. Stuttgart: av edition.
Hague, C., & Jenkins, P. (Eds.). (2005). *Place identity, participation and planning* (Vol. 7). Hove: Psychology Press.
Halegoua, G. R. (2020). *The digital city: Media and the social production of place* (Vol. 4). New York: New York University Press.
Halegoua, G. R., & Polson, E. (2021). Exploring 'digital placemaking'. *Convergence*, *27*(3), 573–578.
Hes, D., & Hernandez-Santin, C. (Eds.). (2020). *Placemaking fundamentals for the built environment*. Singapore: Palgrave Macmillan.
Hespanhol, L., & Tomitsch, M. (2014). Understanding the effects of contextual constraints on performative behaviour in interactive media installations. *Personal and Ubiquitous Computing*, *18*(7), 1651–1665.
Hespanhol, L., Haeusler, M. H., Tomitsch, M., & Tscherteu, G. (2017). *Media architecture compendium: Digital placemaking*. Stuttgart: av edition.
Hodson, M. and McMeekin, A. (2021) Global technology companies and the politics of urban socio-technical imaginaries in the digital age: Processual proxies, Trojan horses and global beachheads'. *Environment and Planning A: Economy and Space*, *53*(6), 1391–1411.
Horton, C. (2018, August 21). The simple but ingenious system Taiwan uses to crowdsource its laws. https://www.technologyreview.com/2018/08/21/240284/the-simple-but-ingenious-system-taiwan-uses-to-crowdsource-its-laws/
Huntington, S. P. (1968/2006). *Political order in changing societies*. New Haven, CT: Yale University Press.
Jacobs, J. (1961). *The death and life of Great American Cities*. New York: Vintage Books.
Kitchin, R. (2022). Conceptualising smart cities. *Urban Research & Practice*, *15*(1), 155–159.
Kitchin, R. (2023). Urban data power: Capitalism, governance, ethics, and justice. *Data power in action: Urban data politics in times of crisis*, 21.
Kitchin, R., Lauriault, T. P., & McArdle, G. (2017). Data and the City. In *Data and the City* (pp. 1–13). London and New York: Routledge.
Latour, B. (2007). Turning around politics: A note on Gerard de Vries' paper. *Social Studies of Science*, *37*(5), 811–820.
Latour, B. (2013). *An inquiry into modes of existence*. Cambridge, UK: Harvard University Press.
Lefebvre, H. (1970/2003). *The urban revolution*. Minneapolis: University of Minnesota Press.

Lefebvre, H. (1991). *The production of space* (D. Nicholson-Smith, Trans.). Oxford: Blackwell.
Lefebvre, H. (1996) *Writings on Cities*. Oxford: Blackwell Publishers Ltd.
Lin, W. I., & Chiu, S. Y. (2019). The mobilisation of creative city building as a new mode of governmentality in Dihua Street neighbourhood, Taipei City. *Geoforum*, *106*, 320–329.
Lippmann, W. (1922/2017). *Public opinion*. London and New York: Routledge.
Mannheim, K. (1937). *Ideology and Utopia: An introduction to the sociology of knowledge*. New York: Harcourt, Brace and Company, 1937), 118.
Markusen, A., & Gadwa, A. (2010). *Creative Placemaking. A white paper for the Mayors' Institute on City Design*. Washington, DC: National Endowment for the Arts.
Mattingly, D. C. (2019). *The art of political control in China*. Cambridge, UK: Cambridge University Press.
McFarlane, C. (2011). Assemblage and critical urbanism. *City*, *15*(2), 204–224.
McQuire, S. (2017). *Geomedia: Networked cities and the future of public space*. Hoboken, NJ: John Wiley & Sons.
Miraftab, F. (2004). Invited and invented spaces of participation: Neoliberal citizenship and feminists' expanded notion of politics. *Wagadu: A Journal of Transnational Women's & Gender Studies*, *1*(1), 3.
Norberg-Schulz, C. (1979). *Genius loci: Towards a phenomenology of architecture*. New York: Rizzoli.
Offenhuber, D., & Ratti, C. (2014). *Decoding the city: Urbanism in the age of big data*. Basel: Birkhäuser.
Özkul, D. (2021). The algorithmic fix: Location intelligence, placemaking, and predictable futures. *Convergence*, *27*(3), 594–608.
Palermo, P. C., & Ponzini, D. (2014). *Place-making and urban development: new challenges for contemporary planning and design*. New York: Routledge.
Pang, C., Neustaedter, C., Moffatt, K., Hennessy, K., & Pan, R. (2020). The role of a location-based city exploration game in digital placemaking. *Behaviour & Information Technology*, *39*(6), 624–647.
Papastergiadias, N. (2006). *Spatial aesthetics: Essays on Art, place and the everyday*. London: Rivers Oram Press.
Pavlićević, D. (2019) *Public participation and state building in China: Case studies from Zhejiang*. London and New York: Routledge.
Peters, M. A., Marginson, S., & Murphy, P. (2009). *Creativity and the global knowledge economy*. Lausanne: Peter Lang.
Polson, E. (2015). A gateway to the global city: Mobile place-making practices by expats. *New Media & Society*, *17*(4), 629–645.
Rajchman, J. (1998). *Constructions*. Cambridge, MA: MIT Press.
Relph, E. (1976). *Place and placelessness*. London: Pion.
Richards, G., & Duif, L. (2019). *Small cities with big dreams: Creative placemaking and branding strategies* (p. 269). Abingdon, UK: Taylor & Francis.
Robbins, S., & Tang, C. (2022 August 22). *Hopes and concerns for Taiwan's new ministry of digital affairs*. https://thediplomat.com/2022/08/hopes-and-concerns-for-taiwans-new-ministry-of-digital-affairs/
Rogelja, I. (2021). The museumification of treasure hill: Authenticity, authority and art in a Taiwanese urban village. *The China Quarterly*, *245*, 227–247.
Sartori, G. (1987). *The Theory of Democracy Revisited*. Chatham, NJ. Chatham House Publishers.
Shelton, T., Zook, M. and Wiig, A. (2015) 'The "actually existing smart city"'. *Cambridge Journal of Regions, Economy and Society*, 8, 13–25.
Simmie, J. M. (1974). *Citizens in conflict: The sociology of town planning*. London: Hutchinson.

Stalder, F. (2018). *The digital condition*. Hoboken, NJ: John Wiley & Sons.
Sweeney, J., Mee, K., McGuirk, P., & Ruming, K. (2018). Assembling placemaking: Making and remaking place in a regenerating city. *Cultural Geographies*, *25*(4), 571–587.
Tang, A. (2019 March 12). *Inside Taiwan's new digital democracy*. https://www.economist.com/open-future/2019/03/12/inside-taiwans-new-digital-democracy
Todaro, D. (2023). "One City, Two Networks": Big Data and Artificial Intelligence for Smart Public Governance in Shanghai. In *Quo Vadis, Sovereignty? New Conceptual and Regulatory Boundaries in the Age of Digital China* (pp. 91–108). Cham: Springer Nature Switzerland.
Toland, A., Christ, M. C., & Worrall, J. (2020). DigitalXPlace. In: D. Hes and C. Hernandez-Santin (eds), *Placemaking Fundamentals for the Built Environment* (pp. 253–274). Singapore: Palgrave Macmillan.
Tomitsch, M. (2016). Communities, spectacles and infrastructures: Three approaches to digital placemaking. In S. Pop, T. Toft, N. Calvillo, & M. Wright (Eds.), *What urban media art can do: Why, when, where, and how*. Stuttgart: av edition.
UN-Habitat China. (2020). Covid-19 Wuhan Guidance Papers. UN-Habitat. https://unhabitat.org/sites/default/files/2020/09/covid-19_wuhan_guidance_papers_2020.pdf
Wang, D. (2003) *Street culture in Chengdu: public space, urban commoners, and local politics, 1870–1930*. Stanford, CA: Stanford University Press.
Wang, W. Y. (2019). *Digital media in urban China: Locating Guangzhou*. Lanham, MD: Rowman & Littlefield International.
Warner, M. (2002). *Publics and counterpublics*. Princeton, NJ: Princeton University Press.
Whyte, W. H. (1980). *The social life of small urban spaces*. New York: Ingram.
Wilken, R., & Goggin, G. (Eds.). (2013). *Mobile technology and place*. London and New York: Routledge.
Willis, K. S., & Aurigi, A. (2017). *Digital and smart cities*. London and New York: Routledge.
Witteborn, S. (2021). Digital placemaking and the datafication of forced migrants. *Convergence*, *27*(3), 637–648.
Xiong, Y. (2022). *Paradigmatic Innovation in Urban Governance: Shanghai's Single Platform for Urban Management*. Beijing: Zhongxin Publishing [in Mandarin].
Zeng, X. Z. (2007). *Community development in Taiwan* [in Mandarin]. Taipei: Yuanzu Cultural Pty Ltd.
Zhao, F., Fashola, O. I., Olarewaju, T. I., & Onwumere, I. (2021). Smart city research: A holistic and state-of-the-art literature review. *Cities*, *119*, 103406.

3 Activist Assemblage

3.1 Introduction: 'Full Stack' Activists

Two decades after Manuel Castells' seminal *Networks of Hope and Outrage*, pan-global social protests either subsided during the Covid-19 pandemic, as seen with the Arab Springs and the Occupy Movement, or became integral to urban landscapes, exemplified by movements like School Strike 4 Climate or Free Gaza. Today, a myriad of localised, self-organised protest actions, seamlessly interwoven through autonomous digital communication networks and technologies, have become ubiquitous, rarely garnering international media attention.

Among these urban protests, a burgeoning civic trend has emerged to counter the urban redevelopment and digital transformation schemes spearheaded by major tech companies and global brands. A prominent instance was the public outcry against transforming the Quayside waterfront into a smart precinct—a collaboration between the City of Toronto and Sidewalk Labs, Google's urban planning and infrastructure subsidiary. The proposed redevelopment has incited fears of privatising public spaces by ceding technological control and data collection rights to private entities, raising profound concerns about surveillance, data privacy and the future of organisational transparency and accountability. Citizens were apprehensive about living in a surveillance city where Sidewalk Labs could capture, analyse and potentially exploit granular personal data for decision-making as well as provide solely privatised public services.

Another noteworthy example is the 'Fuck Off Google' campaign in Berlin, where a neighbourhood vehemently opposed the establishment of a 'Google for Startups Campus'—a co-working space and start-up hub endorsed by the government. Beyond fears of gentrification, with startups displacing local businesses and inflating rents, protesters took issue with Google's stances on public matters such as taxation, data collection and the overarching 'Silicon Valley ideology'. The protest manifested through flyers, zines, posters, talks, demonstrations, an 'anti-Google Café' in an anarchist bookshop and a monthly 'Noise Against Google' protest, where residents made a cacophony with pans, pots and whistles. Similarly, public resistance thwarted Amazon's

DOI: 10.4324/9781003407973-3

proposed headquarters in New York City, driven by gentrification fears and debates surrounding a new Californian city conceived by Silicon Valley moguls. In China, protests adopt more subtle forms, such as artist Deng Yufeng's performance art project 'A Disappeared Movement'. In one act, Deng and a team of ten donned high-visibility jackets and moved in eccentric patterns down a bustling Beijing street, zigzagging, dodging, crouching, creeping and evading the city's extensive surveillance network.

The existing narrative fails to unify these seemingly disparate events and their implications. I interpret these events as a form of activist digital placemaking that diverges from concepts like guerrilla urbanism (Hou, 2010) for three primary reasons. First, these activities are not confined to urban peripheries but often occupy central stages. Second, the activism focuses not solely on survival or mundane urban issues such as rising rents and gentrification but also on abstract conditions influencing human co-existence, such as the privatisation of public spaces, aesthetics, privacy and democracy. Most importantly, these actions are propelled by what I term the activist assemblage: defiant networks of individuals and artefacts at the confluence of city, technology and politics, actively and persistently reshaping the digital city's landscape.

The activist assemblage embodies a 'full stack' quality. Possessing profound socio-technical skills and extensive social networks, it commands the conviction and expertise to mobilise a spectrum of technical, cultural, legal, regulatory and affective resources, manipulating various levers to create favourable outcomes. More significantly, the activist assemblage offers an inclusive framework for conceptualising a new type of alliance. It encompasses all those with an activist mindset and all actions contributing to activist causes. A politician dissenting from the majority opinion, a satirical cartoon, the acts of commenting and retweeting, organic or organised street protests, or even a Molotov cocktail in extreme cases all form part of the activist assemblage as they catalyse change through a series of chain reactions for better or worse.[1] This inclusive model engenders hope by lowering the threshold for participation in urban polity. It requires no formal involvement in an occupation movement but a fundamental level of political awareness, curiosity and empathy as opposed to ignorance, cynicism and apathy. The expanded notion of the activist assemblage also serves as a cautionary tale, urging all participants to be accountable for their words and actions, to be savvy about their resources, means and ends, and to be mindful of the risks and unintended consequences.

I analyse Federation Square's activist assemblage in this chapter, focusing on the 'Our City, Our Square' campaign. By unpacking how the Apple plan was revoked by two of Federation Square's powerful governing and managing bodies, I structure the findings around three phases of the campaign: the formation of the organisation, the development of collective responses and the culmination of the campaign against the Apple Federation Square proposal. Additionally, I compare this campaign with Everybody's East Lake, an urban intervention in Wuhan, China. This chapter underscores how activist digital placemaking can drive urban intervention in redevelopment processes and blur the lines between formal and informal placemaking. Without romanticising the hardships of urban political struggle or overlooking the creative potential of the activist assemblage to effect tangible urban transformations, this chapter contends that the activist assemblage provides structural resistance within urban systems.

3.2 Citizens Melbourne and Our City, Our Square

3.2.1 The Formation of Citizens for Melbourne

The activist group, Citizens for Melbourne, comprised nine core members, primarily architects, urban historians and media professionals. They were bolstered by over 50 peripheral campaigners, including volunteers, performers, content contributors and workshop organisers. At the peak of their activism, rallies orchestrated by the group drew hundreds of protesters, advocates, political supporters and local media. They launched the Our City, Our Square campaign to oppose tech giant Apple's proposal to establish a flagship store in Federation Square. Their website (https://www.ourcityoursquare.org/) and Instagram account (@ourfedsquare) have been active since January 2018. This campaign lasted 18 months until Apple discontinued its plans. After that, Citizens for Melbourne continued to act on other concerns relating to Melbourne public spaces.

The core group was assembled by a heritage-conscious architect, Rohan Story, who dashed to mobilise opponents of the 'Apple Federation Square' concept through public forums. He enlisted petition starters against the Apple plan on Change.org, Brett de Hoedt and Melinda Ovens, who subsequently became vice president and core member of Citizens for Melbourne. Vocal contributors to local news outlets also captured his attention. For instance, Tania Davidge, president of the Citizens for Melbourne, penned an early commentary to ArchitectureAU, while James Lesh, an urban historian and academic, published an opinion piece in *The Age*, a Melbourne daily newspaper. To formalise their operations and lend credibility, the group swiftly registered as an official fundraising organisation, Citizens for Melbourne Inc., with designated roles of president, secretary and treasurer. Citizens for Melbourne identified a niche in their value proposition and distinguished itself as one of the few groups in Melbourne advocating for contemporary public space issues. They operated on a consensus basis, with decisions made collaboratively through a deliberative process overseen by the president, who avoided unilateral decisions.

Interviews with core and peripheral activists revealed that members possessed a diverse array of 'socio-technical skills' prior to the Our City, Our Square campaign. Beyond a shared interest in urban public space, placemaking and heritage conservation, many were already engaged in digital placemaking through public art projects, blogging and digital photography activism. For example, Tania Davidge, an architect then pursuing a PhD at the University of Melbourne, co-founded the non-profit OpenHAUS (www.openhaus.org) to engage the public in architectural dialogues through participatory art projects and discussion forums, revitalising neglected spaces with light installations and using Photoshop to re-imagine public spaces with young people.

John Jović, a former photographer, advocated for heritage protection by documenting the disappearing local environment. Since 2007, he has managed the Instagram account @goneforeverproject and another on brutalist architecture, @brutalistarchitecture. He explained that @goneforeverproject began as an initiative to create triptychs of specific houses or buildings facing demolition, culminating in exhibitions that juxtaposed the original structure, the vacant lot, and the subsequent development. Similarly, Lesh, with a PhD from the University of Melbourne, was an urban and cultural heritage researcher and consultant. Lesh maintained a blog named Heritage City, where he shared insights on urban heritage conservation with a broader public audience.

3.2.2 *The Our City, Our Square Campaign*

The Our City, Our Square campaign deployed a sophisticated hybrid strategy intertwining conventional activism (e.g., direct political engagement through public petitions, submissions and presentations at public hearings) with performative elements (e.g., organising street theatre and creating visually compelling images for social media dissemination). The campaign seamlessly navigated both online and offline spheres, employing a diverse array of activities from public forums, rallies and crowdfunding initiatives to Instagram hashtag campaigns and multi-channel media releases.

Three key campaign features can be discerned. First, the pervasive use of digital technologies was integral, fostering a sense of connection and mobilisation among the audience. Networked platforms like Twitter, Facebook groups and email lists served the purposes of connection and mobilisation. For instance, leveraging an email list of 4,500, the campaign galvanised 2,360 public submissions urging Heritage Victoria to reject the Apple proposal (Davidge, Interview, 1 May 2019). The campaign website functioned as a repository for campaign narratives and action updates, occasionally retailing merchandise to offset operation costs. It outlined seven compelling reasons to support the campaign and presented a five-point People's Plan for Federation Square. Satirically dubbed the '#OurCityOurSquare flagship store', the website also sold T-shirts and tote bags. To disseminate content and generate momentum, Instagram and Twitter proved invaluable, with @ourfedsquare producing over 200 posts and garnering more than 1,000 posts under hashtags such as #ourcityoursquare, #fedsquarenotapplesquare and #ourfedsquare.

Second, the campaign operated on a volunteer-based, low-budget model. In the absence of sponsorships, the campaign's materials were economically designed, exemplified by posters printed with black ink on fluorescent paper, the most cost-effective method for creating impactful visuals (Davidge, Interview, 1 May 2019). Budget constraints challenged the campaign's sustainability and volunteer motivation. The considerable affective labour involved in activism often leads to attrition over time, necessitating strategic allocation of limited resources. This included maximising social media engagement and fostering relationships with local news outlets to secure coverage. Lastly, the campaign emphasised clear and relatable messaging. Given the abstract nature of public space, particularly to the untrained public, the campaign focused on delivering straightforward, concrete messages to help the public visualise the proposal and its implications (Davidge, Interview, 1 May 2019). Architectural jargons and theoretical discussions of urban public spaces were minimised. Throughout the campaign, cultural and political references from local and global events were frequently appropriated to enhance relatability.

3.3 Contentions

3.3.1 Secretive Urban Decision-making

At the heart of the campaign, public debates converged on several major contentions. Many activists recalled early rumours of the Apple proposal circulating a year before its official announcement. Those who disregarded the rumour at that time were astonished by its abrupt actualisation without any public notification or consultation. In Australia, community engagement for city planning is typically mandated by local and state governments (Stewart & Lithgow, 2015), yet exceptions can be engineered through bureaucratic processes. To fast-track the Apple deal, Richard Wynne, the then Minister for Planning, bypassed the need for public notice and exempted Apple from applying for a planning permit under Clause 52.03 of the Victorian planning scheme (Novacevski, 2017), thereby enabling Apple to pursue a parallel planning process.

Most local councillors from the City of Melbourne (COM) were informed only post-announcement.[2] For example, Rohan Leppert, local councillor and supporter of the Our City, Our Square campaign, articulated in the interview that the clandestine nature of the Apple plan was a deal-breaker. 'The fact that it was secret and announced a week before Christmas was in and of itself enough to kill the entire thing' (Leppert, Interview, 11 June 2019). Activists perceived the timing of the announcement as a typical governmental stratagem to surreptitiously introduce contentious agendas into public discourse during a period of general distraction.

The opacity enveloped the legality and financial ramifications of Federation Square Proprietary Limited's (FSPL's) agreement with Apple. The terms and conditions of the lease were withheld from the public under commercial-in-confidence provisions. The Victorian Government declined to divulge any financial specifics of the Apple plan, merely assuring Victorians that they would not bear the costs. The lack of transparency regarding Apple's management of the site incited concerns about the commercialisation of Federation Square, potentially compromising its integrity and undermining its public and civic essence.

3.3.2 Privatisation and Commercialisation of Public Space

The Apple proposal epitomised a neoliberal agenda, prioritising profit over the preservation of urban public spaces. Rather than integrating into the original site aesthetics, the proposed Apple town square threatened to overshadow it, effectively rebranding Federation Square as a quasi-public space dominated by the Apple logo. The debate over financing public spaces has pervaded Federation Square's history. Lesh critiqued the notion of profitable public spaces, a neoliberal trend that gained traction in the 1970s US and that subordinated all else to market forces (Lesh, Interview, 9 May 2021). Activists contended that the 'town square' model may be deceptive and not as inclusive as it advertised. For instance, participation in Apple workshops required an Apple ID and, ideally, the latest Apple product.[3] This condition meant that the promised public educational courses were restricted to Apple's current and prospective customers. Davidge (Interview, 1 May 2019) shared an architectural rendering of Apple Kungsträdgården in Stockholm, noting how the design imposed an amphitheatre with a glowing wall, a waterfall, and the Apple logo as an 'Instagrammable' backdrop for selfies.

3.3.3 Site Integrity and Cultural Heritage

Initially polarising, the aesthetics of Federation Square, gradually gained acceptance among Melbourne's public. Jović, for instance, championed not a beloved building but the site's integrity. Though initially indifferent to Federation Square's design, he grew to appreciate it. 'I really love it. I was walking past there the day after the public hearing, and the light was just right, at the right angle, at the right time of day. And it's actually just a stunning place' (Jović, Interview, 23 April 2019). This sentiment echoed Swiss architect Peter Zumthor's (2008) concept of 'architectural atmosphere', by which he emphasised that the sense of place is achieved beyond mere delivering architectural materialities, but through the arrangements of phenomenological elements such as light, angle, weather, mood and social context. The unique socio-material configurations of Federation Square also spurred discussions about its heritage conservation among many activists. As places are fluid and socially constructed through human interactions, 'heritage is about systematically managing inevitable changes' (Lesh, Interview, 9 May 2021). Designating a heritage site involved comprehensively understanding Federation Square's architectural, social and historical value with a forward-looking perspective.

3.3.4 *'Not about Apple'*

For my interviewees, the campaign was both 'about and not about' Apple. The global tech giant naturally became a focal point of controversy, leading to divergent content strategies within the campaign. One contributor to the media release insisted that the campaign was not anti-Apple and should refrain from slogans like 'Stop Apple,' 'Block Apple' or 'No Apple at Fed Square.' Instead, the key message emphasised positively asserting the community's right to the space. Conversely, Brett de Hoedt, a media professional managing the campaign's media and public relations, recognised Apple's evocative power to generate widespread media coverage. This dichotomy resulted in two sets of campaign slogans and hashtags: #FedSquareNotAppleSquare and #OurCityOurSquare.

Concerns surrounding the commercial implications of Apple's proposal were palpable among activists. One conceded that opposition might have been unnecessary if the Yarra Building's aesthetics and usability remained unchanged. Yet there was suspicion that Apple Federation Square would impede public rights to protest in a public space, since Apple may hinder FSPL's allowance of civic activities deemed to obstruct its normal commercial activities under a lease agreement. Some activists perceived that Apple's socio-spatial symbolism and influence on urban public spaces diverged significantly from traditional commercial or state-run initiatives. For Davidge, Apple's commercial attributes and socio-spatial impact contrasted starkly with those of existing businesses and state-owned enterprises at Federation Square. Similarly, Leppert argued that the Apple store could undermine the Square's character, transforming it from a nondescript public space into a commercial one, asserting that 'the closest you get to corporate branding [on the square] is Special Broadcasting Service (SBS), but people aren't going to feel the same way about SBS as they do about a purely profit-driven international company' (Leppert, Interview, 11 June 2019).

Despite Apple not relying as heavily on surveillance capitalism as the likes of Google or Facebook, and privacy breaches being a lesser concern for activists, the concept of Apple Federation Square echoed the notion of incursion in Shoshana Zuboff's (2019) 'dispossession cycle'. Incursion involves the virtual ability of the dispossessor to claim undefended domains and extract surplus value from the nonmarket spaces of everyday life. Similarly, Apple Federation Square sought to appropriate the unique civic and cultural value of Melbourne's central meeting place—unrecognised and unmeasured by the Victorian Government—transforming it into the purchasing power and branding effect.

This scenario substantiates Latour's assertion that concerns generate publics. Matters of fact are risk-free and objective, whereas matters of concern encapsulate complexity, incite agitation and provoke public discourse (Latour, 2004). Various concerns galvanised multiple publics—designers and architects emphasised aesthetic integrity, heritage experts focused on social, cultural and heritage values, and media professionals addressed communication strategies and public relations. United by a love for public spaces, these groups felt compelled to advance the discussion. This dynamic resonates with digital placemaking as a heterogeneous and relational field where multiple concerns emerge, evolve, resolve or fade.

3.4 Intervention

The campaign intervention can be analysed into two categories: what I term performative and design intervention. Here, my usage of 'performative' and 'design' originates from Boris Groys (2014), who pinpointed two political tendencies of art activism. Artistic aestheticisation accepts the status quo as dysfunctional by critiquing reality, while design aestheticisation proposes to change the status quo by rewiring reality. This distinction aligns with my understanding of Our City, Our Square's intervention practices. Performative intervention leveraged the urban stage to enrich the existing urban symbolism by reframing, reimagining and renegotiating the identity of Federation Square. Design interventions engaged directly in urban political and decision-making processes and aimed to achieve specific objectives regarding Federation Square's future. The two modes are distinct yet connected.

3.4.1 Performative Intervention

Urban activism employed visual politics and affective performativity in tandem for performative intervention. Given the abstract nature of public space, architectural renderings facilitated public imagination and rehearsed a sense of loss. Citizens for Melbourne produced architectural renderings of the Apple store on Federation Square to visualise the potential change, superimposing human silhouettes against the Apple design to caution locals and tourists about becoming faceless Apple consumers. Early Apple advertisements were reappropriated: for instance, the video 'Just another reason to switch to iPhone. iPhone vs your Phone' (FahadThetics, 2017)—originally meant to illustrate the ease of transferring music from 'your phone' to an iPhone—was repurposed to propose alternative locations for Apple's flagship store, suggesting a move from Federation Square to Docklands. This video, provided by the anonymous contributor @stopappleinc, exemplified the campaign's innovative approach. Despite pressures to maintain a consistent presence, activists exercised restraint, selectively screening content, contrasting with the more explicit protests against Google's Berlin Campus. For example, the campaign's appropriation of Apple's iconic 1984 Superbowl advertisement, altering 'Think Different' to 'Think Demolish' with the voiceover 'Here's to the crazy ones,' was swiftly removed from Instagram because of its overly destructive aesthetics and negative tone towards Apple.

Fluorescent posters visually branded the campaign in physical gatherings, Melbourne's alleyways and online environments. Budget constraints necessitated minimal yet impactful designs, featuring black print on neon-coloured papers.

The group's strategic timing and engagement with current local affairs were astute. Lesh recognised the necessity of being distinct and promptly responsive to new developments, referencing high-profile cultural and social events such as the Sydney Opera House controversy and the COM's Lord Mayor by-election. Being quick and going local would help to keep the campaign's momentum. Swift and localised responses sustained the campaign's momentum. Interventions such as architectural renderings, appropriated advertisements and fluorescent posters resonate with Elwood and Hawkins' (2017) assertion that visual politics obfuscates visual symbols and reflects power shifts in visual regimes. The campaign contributes to a networked production of new urban imaginaries by recreating urban images and reclaiming urban spectatorship.

Affective performativity manifests vividly in a pop-up street theatre and multiple public rallies. The street theatre production, titled 'Poison Apple,' took place on Federation Square. It commenced with 20 individuals each biting into a poisoned apple and dramatically collapsing. An actor, costumed as an apple, emerged guiltily as the antagonist. Subsequently, a procession comprising bananas, grapes and oranges stormed in, expelling the apple from the square, chanting, 'You don't belong here, Apple!' The enchanted audience then revived and celebrated by eagerly consuming their apples while chanting in unison, 'Juice the apple, juice, juice the apple.' This collective emotion reverberated to uninformed passers-by, many of whom paused to observe and film the performance, albeit without fully grasping its meaning. For instance, a tour guide, seemingly non-English-speaking, danced to the chant's rhythm, waving his fist and mumbling words akin to 'tree tree apple,' followed by a trail of wide-eyed tourists.

Analogous approaches were employed during several public rallies. On one occasion, the presenter invited a girl on stage to ironically question her iPhone, 'Hi, Siri. Where is a better place in Melbourne for Apple to build its flagship store?' Siri politely responded, 'I don't understand what you meant'. On another occasion, the group announced a crowdfunding event aiming to raise $40 million to purchase Federation Square. Many symbolic gestures fostered a sense of playfulness and conviviality. The tongue-in-cheek crowdfunding campaign came out of a conversation about the Sydney Opera House. 'How much would each person in Sydney be willing to spend dollars per year to keep Sydney Opera House a public venue, right', asked the activists, 'what about Fed Square?' This dialogue evolved into a campaign aimed at garnering global media attention (Lesh, Interview, 9 May 2021). The street theatre, public rally and announced crowdfunding all echoed Duff's (2017) notion of 'the affective right to the city', materialising through the social and affective elements of mediated performance.

3.4.2 *Design Intervention*

Citizens for Melbourne engaged directly in the decision-making processes regarding Federation Square. Davidge (2019) documented the wide range of activities undertaken in the campaign's annual review. Citizens for Melbourne interfaced with local media, heritage activists and consultancies and organised protest rallies. Notably, a rally attended by around 200 individuals on a frigid Wednesday evening in September featured prominent speakers such as Simon Ambrose, CEO of the National Trust Australia (Victoria), Colleen Peterson from Ratio Planning Consultants, and COM Councillor Rohan Leppert. The event was covered by major local press, including Channel 7, Channel 9, *The Herald Sun* and *The Age*. Peter Davidson, a key architect of Federation Square who opposed the Apple plan, attended despite being unable to speak publicly due to his medical condition and contributed one of his artworks for T-shirt fundraising.

Moreover, the activists engaged with politicians either supporting or opposing the Apple plan, orchestrating meetings and conversations and facilitating over 600 emails to state politicians in opposition (Davidge, 2019). Despite its non-partisan stance, the campaign sought political backing and was visibly active in local and state elections. In response to the mayoral by-election coinciding with the Apple plan announcement, the activists launched a poster campaign, inviting candidates to be interviewed about their stance on the Apple Flagship. They later disseminated an alternative voting card based on the candidates' responses, thus enhancing exposure for supportive candidates and drawing media and local attention to their cause.

Additionally, Citizens for Melbourne played a crucial role in the process of listing Federation Square on the Victorian Heritage Register as a place of state-level cultural heritage significance. In December 2018, they submitted their support to the Heritage Council of Victoria and provided a *pro forma* submission on their website, facilitating over 2,100 applications. Despite being denied access to public submissions without a formal Freedom of Information request,[4] I was informed that many applications were based on the *pro forma* available on the Our City, Our Square website. Consequently, Citizens for Melbourne were invited to present at the public hearing ordered by the Victorian Heritage Council.

To summarise, the campaign underscores a network enabling urban placemaking interventions. A critical aspect of this campaign is the integration of performance, activism and digital technologies in urban space. All performances were recorded, uploaded, curated and shared online, amplified through email lists, websites, Instagram accounts and hashtags. Papacharissi's (2015) concept of 'affective public' mediated by social network technologies highlights how these can destabilise power hierarchies and foster communal solidarity, meriting recognition in contemporary politics often grounded in rationality. The campaign echoes Duff's (2017) advocacy for shifting from a juridical to an affective conception of Lefebvre's 'right to the city', accommodating its performative expression. The campaign reaffirmed that the networked technologies in public activism are not merely digital instruments but reflect 'a grounded and politically-oriented use' (Puente, 2011, p. 343) in the increasingly hybrid urban space.

3.5 Activist Digital Placemaking

3.5.1 Our City, Our Square as Activist Digital Placemaking

Possessing digital, performative and political components, the Our City, Our Square campaign exemplifies activist digital placemaking. The campaign illuminated how public gestures infiltrated multiple placemaking processes, ranging from the symbolic realms of urban imaginaries to practices of dwelling and inhabitation, thereby materially transforming place programming, decision-making and policies. This movement unveiled the unofficial public history and cultural significance of Federation Square, which the state government and FSPL had undervalued. Through potent performative and affective elements, the campaign underscored Federation Square's identity as fluid rather than static, constructed rather than definitive and generative rather than prescriptive. Rancière (2003) broadened the concept of political action to encompass the (re)configuration of urban space, the (re)thinking of existing urban parameters and the (re)framing of urban experience. This expanded definition acknowledged a 'sense of politics' inherent to the public, often excluded by parliamentary and party politics. The confluence of public-making and activism emerged as a 'natural' response to the Apple controversy, with the campaign's strategy resonating with what Jacobs (1961) termed the 'spontaneous self-organisation of urbanism' throughout extended periods of urban life. The status quo of urban governance was disrupted and reconfigured, positioning the campaign as a generative force that underpinned the interdependence of urban imaginaries and built environments.

The Our City, Our Square campaign embodies three core characteristics of the activist assemblage. First, advancing the discourse of activists as disobedient and networked, the assemblage comprises diverse political actors armed with exceptional socio-technical skills. While prominent figures in urban politics include architects, designers, planners and urbanists, the assemblage also encompasses media practitioners, artists, journalists, politicians, culturalists, accountants and various other professionals. Many of these individuals pursue multiple careers (Alboher, 2007) to achieve work/life success and fulfil creative needs. For instance, many activists are also active bloggers and content creators within their fields. Actions, in Hannah Arendt's sense, that render oneself visible and audible in the public realm—such as blogging, Instagramming, participating in creative festivals and writing for magazines—subsequently enhance the activists' confidence and skills, enabling further contributions to urban public affairs.

Drawing upon the five predominant skills of 'technopolitics' or 'nerd politics' (Postill, 2018), I identify the critical socio-technical competencies for the activist assemblage as encompassing legal, artistic, media and entrepreneurial proficiencies. Mastery of the legal parameters of land use and planning schemes in urban development is paramount. Artistic abilities, such as digital storytelling and pop-up urban theatre, are vital for visualising proposed urban transformations and articulating latent collective emotions as compelling cultural narratives in an increasingly ocular-centric communicative era. Activists adeptly navigate both analogue and digital media landscapes. They leverage digital and networked technologies—including email lists, social media groups, dedicated campaign websites, Twitter and Instagram—to mobilise campaigns and disseminate content, thereby amplifying the campaign's core message across various platforms. Building relationships and trust with public media outlets, such as newspapers, radio and television, has resulted in proactive media engagement as the campaign gained momentum and social license.

Entrepreneurial skills, often overlooked in the context of activism, are essential. Given the volunteer-based, low-budget nature of urban activism, sustaining long-term engagement, motivating volunteers and rewarding their physical and affective labour pose significant challenges. Activists require entrepreneurial acumen to judiciously allocate limited resources, communicate empathetically, mediate diverse relationships across multiple networks and occasionally subsidise campaign costs through fundraising. For instance, the campaign website, in addition to documenting campaign visions and activities, periodically sold merchandise like T-shirts and tote bags through a page whimsically named the '#OurCityOurSquare flagship store.'

Moreover, the campaign illustrated that the activist assemblage is deeply intertwined with other urban layers and networks. It was spearheaded and sustained by a loosely connected coalition of skilled individuals and technical entities operating within a liberal democratic framework. This coalition included the COM Council, National Trust of Victoria, Victorian Heritage Council, local media outlets, the general public and their email lists, both identified and anonymous contributors to Instagram accounts, and passers-by on social media platforms and mobile devices. The campaign renegotiated Federation Square's identities, requiring endorsement and legitimisation by traditional institutions within the existing power paradigm. This assemblage exemplifies the interplay between the prescriptive power of traditional institutions and the generative power of activists, thereby blurring the lines between informal and formal placemaking. It serves as a real-life response to Brenner's (2020) assertion, demonstrating that informal placemaking can indeed precipitate political change.

3.5.2 Activist Reflections

The activist and their collaborators positioned themselves as the intermediaries for channelling community outrage. They actively utilised an amalgamation of professional, social and virtual networks, amplifying these channels with interpersonal interventions. For instance, when Peter Davidson donated one of his architectural designs to Citizens for Melbourne for T-shirt production, he not only endorsed the campaign personally but also broadened its reach and impact. Similarly, cartoonist Matt Golding's contribution of political cartoons fortified the public value against Apple's commercial interests, transferring the sentiment to a broader audience. These disparate yet personal acts coalesced into a resilient, networked campaign potent enough to instigate social change. Unlike traditional advocacy groups such as the National Trust (Victoria), Citizens for Melbourne exhibited a more fluid and adaptive operational style, responding swiftly and decisively compared with the conventional deliberative approach. Although many members of Citizens for Melbourne possessed substantial socio-cultural capital, their motivation appeared to stem from genuine 'interest' and 'affection' for urban public spaces, particularly Federation Square. These emergent public interests challenge the rigid dichotomy between elites (top-down experts and practitioners) and grassroots (organic public expressions). Consequently, Our City, Our Square diverged significantly from the 1997 heritage campaign against Federation Square's shards organised by the National Trust (Victoria) (Lesh, 2021). The campaign epitomised Bennett and Segerberg's (2013) notion that networked individuals with diverse affiliations have become pivotal to activist processes, superseding traditional social organisations.

Ultimately, the campaign underscored that activist assemblages operate within a more dynamic and multilayered urban political landscape than previously envisaged. Celebrated as tech-savvy and well-organised agents driving a decentralisation of power, activists grappled with the intricacies of advocacy and urban politics. Reflecting on the power shifts facilitated by the campaign, Leppert acknowledged that it had endowed the COM with greater leverage in negotiations concerning Federation Square. The alignment of the COM with public sentiment had transformed the political landscape. Leppert noted that, prior to the campaign, decisions regarding Federation Square were confined to the Victorian State Cabinet, excluding the COM. However, post-campaign, COM representative Rob Adams[5] joined the State Federation Square Review panel, signifying an enhanced role for the COM in conveying local opinion to the Victorian Government.

Activists further contemplated the complexities of urban politics and their campaign's impact. For Lesh, heritage advocacy was interwoven with the multifaceted trajectories of urban complexities: inquiries into urban development, societal structures, economic considerations and broader political contexts. Whenever attempts were made to extricate heritage from its milieu, they invariably failed because heritage consistently intersects with concurrent phenomena. 'Our heritage advocacy was similarly intertwined, creating a convergence where distinguishing heritage advocacy from broader civic engagement became challenging. Critics often accused us of instrumentalising heritage for various agendas, but my interest lay in deconstructing these overlapping complexities' (Lesh, Interview, 9 May 2021).

Davidge remarked on the intricate nature of urban politics, revealing it to be far less straightforward than initially perceived. First, a permanent cause or steadfast alliance did not exist. Federation Square, a vestige of Liberal Premier Jeff Kennett's tenure,[6] did not garner Liberal support in conjunction with the Greens to safeguard its legacy. Second, discerning a 'correct argument' proved elusive to other parties. Reflecting on her inaugural foray into activism,

Davidge confessed: 'I thought, we've got all the right arguments; surely everyone will just see the light, and it will all be fine. Then you realise that politics is more complex and blunt than that' (Davidge, Interview, 1 May 2019).

For activists, the complexity of urban politics materialised through a mosaic of trajectories, stances and alliances. Politics manifests not as a logical debate delineating right from wrong but as a tapestry of multilateral relations wherein all participants possess legitimate grounds for their actions. The notion of politics as an absolute dichotomy between irreconcilable adversaries can be renewed by acknowledging the confluence of diverse concerns, practices, beliefs and contingencies.

3.5.3 Researcher Reflections

I reflect a need for further dialogue. First, while ostensibly focused on community placemaking, the campaign arguably struggled to engage a broader, multicultural audience. An interviewee noted that most activists were university-educated professionals with industry ties. Early campaign efforts exhibited a degree of inexperience; for instance, the street theatre's purpose was not explained, leading passers-by to mistake performers for 'local fruit promoters'. Additionally, Indigenous and multicultural participation was sparse at rallies and public hearings. However, this lack of diversity cannot be solely attributed to the activists, who were receptive to my interview requests and follow-up discussions.

My analysis of the activist assemblage underscores the imperative for deeper scrutiny into the socio-spatial ramifications of tech corporations in urban spaces. Activists critically reflected on the digital transformation of urban public spaces, revealing nuanced perspectives. Activists concurred that public spaces often mirror broader societal structures and necessitate digital capital for their success. However, achieving equilibrium and fostering deliberation were deemed essential. Lesh emphasised that urban planners should strategically incorporate technological elements to reinforce the public mission of these spaces. Conversely, Davidge advocated for a context-specific approach to technological programming in urban spaces:

> How might the technology be used in that space? In different spaces in the city, it'll be a different thing. But you need to think about what your aims and objectives are ... The Apple store always pinned this kind of idea that it was going to be innovative without really giving us an understanding of what 'innovative' is. And I'm always a little bit wary of the word innovation, because it's often used to plaster over some other cracks. ... It's like, are we talking about using screens at Fed Square or are we talking about just technology in general? Or are we talking about Apple as a technology provider? Like there's different all these different layers and different ways you can cut it.
>
> (Davidge, Interview, 1 May 2019)

Techno-capitalism's socio-spatial effects on urban public space appeared as a question that begs further discussion. Suarez-Villa (2012, 2021) has portrayed techno-capitalism as a new mode of capitalism in which corporate organisations are designed to exploit the intangibles, such as creativity, knowledge and brand value. Retail consultant Doug Stephens (2017) has argued that retailers should shift from thinking of their stores as distribution channels to utilising them as media channels that deliver brand-building and emotionally compelling experiences to the customers, which may not result in immediate transactions. Apple recognised this opportunity, took a step further and pivoted to deliver a combination of innovation, community and public culture. The public response revealed a significant and visceral level of distrust regarding whether a tech corporate can assume the function of delivering public culture, which is usually associated with the government.

3.6 Beyond Australia: Everyone's East Lake in Wuhan, China

Do activist assemblages in disparate locales and political contexts similarly provide resistant buffers against urban processes? I will revisit an unsuccessful case of urban activism in Wuhan, China, through the lens of assemblage theory[7] and extrapolate its broader implications. In March 2010, the Overseas Chinese Town (OCT) Group, a prominent Chinese urban developer, unveiled plans for a substantial investment in a luxury residential district featuring a 'Happy Valley' theme park and a resort hotel in Wuhan's East Lake Scenic Area. The Wuhan government, anticipating a boost in tourism, endorsed the project and sanctioned OCT's proposal to infill portions of the lake and displace adjacent villagers and fishery workers. This development incited public dissent, eliciting critiques concerning the degradation of water quality, privatisation of public lake access and processes of gentrification (Everyone's East Lake, 2010)

As a response, architect Li Juchuan and artist Li Yu launched the 'Everybody's East Lake' art series in 2010 when the project broke ground, in 2012 during the grand opening of Happy Valley and in 2014. The initiative intended to open up the discussion in an artistic way, exercise new modes of urban occupation and create opportunities for people concerned about the commercial development and the future of East Lake to express their opinions. An official website, 'Everyone's East Lake' (http://donghu2010.org/), has been set up and encouraged everyone to create and implement an artwork about East Lake and report the detailed accounts and information back to the website. The initiative attracted hundreds of participants, ranging from artists, architects, designers, punk musicians, theatre workers, poets, unemployed people and students to scholars and programmers.

Similar to the Our Fed Square campaign, hybrid artistic interventions have been critical to the urban intervention of Everyone's East Lake, albeit in a much more elaborate manner. For example, the 2010 East Lake for Everyone (Round 1) has assembled 53 individuals and groups and produced 59 artworks.

A group of artists occupied physically (and virtually on Google Maps) the remains of a wall from an old fishery dormitory 33.7 meters from the lake to mock the government's arbitrary 30-meter greenbelt rule scrambled after the announcement, contrasting it with the 100-meter buffer zone recommended by the ecologists. Swiss artist Yovski has hosted an event themed around 'a so-called mental patient's counter-diagnosis of a so-called pragmatic era', linking the 'counter-diagnosis' to the critique of capitalist rationality behind the destruction of East Lake. Artist Zijie and friends organised a live-action role-playing game of a scavenger hunt around the East Lake. Other performative activities include producing audio-visual documentation of the East Lake, touring and temporarily living around the lake, drinking and selling water from the lake, contaminating the East Lake with a drop of artist's blood and sending a text message 'Everyone's Donghu' Art Project' to an interactive public screen in Beijing that displays customer messages via SMS.

However, the Everyone's East Lake initiative conspicuously lacks the design intervention. The artists shunned away from formal organisation, partly to avoid political scrutiny and repression and partly to embody their belief in individual autonomy and discursive political resistance. Consequently, the initiative forfeited a leadership role in negotiating overt forms of protest. This absence of structural support left evicted villagers and fishery workers without advocacy in their compensation negotiations with the local government. A planned protest march involving hundreds of urban residents and students was aborted due to police intimidation, school warnings and internet disruptions targeting the organisers. Other forms of petitions, newsletters and public discussions on internet forums were swiftly censored.

In dissecting the failures of the Everyone's East Lake initiative to halt the development and commercialisation of East Lake, conventional wisdom often points to China's stringent public protest controls and pervasive online censorship. However, as Daniel Mattingly (2019) elucidates, attributing the failure solely to repression offers an incomplete picture of the Chinese state's societal control mechanisms. Pure repression is costly and ultimately undermines the regime's popularity (Pavlićević, 2019). The crux lies in civil society's interaction with covert control mechanisms. An assemblage renders operationality and optionality of this instance and reveals the lack of clarity of the core issue, a need for formation and upskilling and a limited capacity for intermediaries to navigate different assemblages within non-democratic contexts.

The core issue of the Everyone's East Lake initiative manifests in a dual sense. First, the primary disputes over the legality and rationale behind the redevelopment have been largely overshadowed by sustainability concerns. The proposed lake filling has incited widespread objections and fervent debates among Wuhan residents on media platforms and social networks. OCT responded with clarifications emphasising the lake filling. However, the crux of the issue, which has been largely overlooked, is the potential illegality of the proposed deal. The Wuhan government leased the nationally protected

East Lake Scenic Area to OCT for redevelopment without securing the necessary permissions from the central government. The entire negotiation and planning process was conducted in secrecy, devoid of public business cases, ecological and social assessments, or any form of consultation.

Moreover, the core contention has been aestheticised and deliberately depoliticised to safeguard the art initiative and its participants. For instance, in the acclaimed art film 'Li Wen at East Lake', director Li Luo blurred the boundary between documentary and magical realism, leading audiences on a journey along East Lake in search of a psychiatric patient, encountering a blend of real and fictional characters, including the director's friends, police, public officials, strangers and historical and mythical figures. Despite its poetic delivery, the film failed to resonate emotionally with a significant cohort absent from the initiative, as detailed below.

The initiative's formation lacked the crucial involvement of the most affected parties: the evicted villagers and displaced fishery workers, who received meagre compensation for relocating during this nominal land deal. OCT allegedly not only acquired the 'use-rights' to the land from the municipal construction company affiliated with the Wuhan government but also appropriated the 'property title' (Yao, 2010). The forcibly relocated villagers lost their entitlement to their previously collective (albeit effectively state-owned) land. This group's voices, imaginations and aspirations for the place, along with their connections to broader natural and social environments, have been neither visualised nor mobilised. Some participating artists (Poxiansheng, 2012) retrospectively reflected on their inability to effectively communicate with the disempowered villagers, who had limited literacy and artistic understanding. These villagers tended to 'play it safe,' desiring no more than external intervention to help increase land compensation and expose the violence and corruption during their forced removal. Despite the villagers being a heterogeneous group with multifaceted needs in complex processes of collective decision-making (Mattingly, 2019), which sometimes involve violence, corruption, village politics, petitions, protests and online petitions, they remained faceless and voiceless in this instance of the activist assemblage. Their socio-technical skills and connections were rendered immobilised.

The Everyone's East Lake initiative has been stymied in its attempts to forge effective linkages with organisational and governmental assemblages, primarily due to the propensity of non-democratic systems to compartmentalise disparate groups. This segregation ensured that all actors were confined within isolated silos, precluding the formation of unexpected alliances and associations that could drive transformative change. Intermediaries have been scarce, and those that exist exhibited limited capability or willingness to mediate across these assemblages. For instance, the activist assemblage demonstrated a more pronounced artistic agency but failed to engage with the principal themes, processes and actors in this placemaking event. In essence, they did not succeed in establishing and maintaining an 'architecture of protest' (Cowan, 2004) that could deconstruct the intricate 'relations and tectonics of people and ideas' (p. 1). This does not imply that activist assemblages in China possess a lower 'relational quotient' to comprehend and navigate complex processes, relationships and contexts; rather, they lack the enabling environments and practices necessary in real-life scenarios.

The activism against the redevelopment of East Lake notably diverged from other protests in China, which focused primarily on safeguarding private property or individual interests amid urbanisation, land seizures and forced evictions (Shin, 2013). Instead, this campaign aimed to preserve an emerging sense of publicness and an autonomous public order, resisting the encroachment of privatisation and the politicisation of public spaces. Mai (2014), a key participant artist in Everyone's East Lake, described public spaces liberated from state-capital intervention as public ruins—temporary sites where state-capitalism, political surveillance and civic autonomy co-exist in a precarious balance. While this balance is susceptible to being disrupted, bulldozed by political machinery, or co-opted by consumerist symbols and rituals, the significance of such public expressions cannot be overlooked. The defiant action known as Donghu Jump, which originated from this campaign, was registered as a local beer trademark in 2016 and transformed into a ticketed annual event at Happy Valley, stripped of its activist meaning.[8] Yet the existence of such public actions in political contexts where physical protests are suppressed and virtual dissent is censored underscores the fluid formation and enduring presence of activist assemblages. As what Bernard Stiegler (2018) called a politics of negentropy, such movements may be temporarily latent or suppressed, but they resist dissipation and retain the potential to eventuate and surprise, as evidenced by the White Paper protest during the pandemic.

Despite varying degrees of success, both Our City, Our Square and Everyone's East Lake exemplify activist assemblages as structural resistance to urban processes. These initiatives embody placemaking, striving to democratise community-building through meaning-making activities and the formation of place identities (Loopmans et al., 2012). Utilising embodied and digital methods, they broadened public intervention within traditional spatial narratives (Frith & Richter, 2021; Page, 2020). Unlike instances of urban activism that tend to be associated with transient formation and event-based activities, activist assemblage provides structural resistance to the urban systems. Activist assemblage makes people associate with temporary activist architecture that is transient and event-based, but activist assemblages demonstrate political longevity by showcasing power through connection and vibration rather than extraction, invasion or subjugation. While Harvey (2008, p. 33) has argued that urban social activists seek to 'reshape the city in a different image from that put forward by the developers, who are backed by finance, corporate capital and an increasingly entrepreneurially minded local state apparatus', I have demonstrated that collaboration, alliance-making, compromising, mediation and negotiation are inseparable parts of the urban political processes to achieve solutions. The faces of the cities are undoubtedly being shaped by these new socio-technical organisms, during which processes transient, micro-actions accumulate to manifest as long-term macro landscape changes.

3.7 Conclusion

This chapter delved into the Our City, Our Square campaign, emphasising its inception, intervention methodologies and the intricate nature of urban political dynamics. In juxtaposition with other initiatives, Our City, Our Square emerged as a coalition of professionals and the public, unified by a shared commitment to urban spaces and endowed with socio-technical skills and functioning networks. The chapter further situated the sophistication of the campaigns in the complexity of urban political processes. It also reflected on the ambiguous identity of the activists and a need to reposition the focus from the identities of the activist to the positioning of the activist assemblage. In this instance, digital placemaking offers a performative conduit for collective actions and networked connections. The analysis revealed that activist digital placemaking attains sophistication through the integration of performative and design interventions, thereby enriching urban semiotics and engaging with political mechanisms. Activist assemblages offer structural resistance within urban systems, intricately weaving inhabitation, symbolism, and urban politics, even in non-authoritative environments where they harbour latent yet nuanced potentials for collective action.

Notes

1 It is unfortunate that some activist assemblages can be entangled with components of violence. However, criminal acts are beyond the discussion of this book.
2 Except for the former Lord Mayor Robert Doyle, who became embroiled in a separate scandal at the time and resigned before the official Apple Federation Square announcement was made
3 This concern was legit. I signed up for a photography class at Apple Chadstone in Melbourne some years ago, hoping to understand more about the colours and compositions. In reality, I was introduced to the latest model, the iPhone 8, which has brand-new portrait feature.
4 For my PhD project, a freedom of Information request has been unattainable due to the requisite time and cost.
5 Rob Adams is an architect and urban designer, Director of City Design at the COM. He is renowned as the leader of revitalisation of the Melbourne City centre and Postcode 3000 strategy.
6 Jeff Kennett was a Liberal premier of Victoria (1992–1999). Federation Square was one of his high-profile capital works projects.
7 For a more detailed account of the 'Everybody's East Lake' project, see Chuang (2022) and Li (2022). For a 'conventional' comparison between the Federation Square and the Donghu cases, see Xiao and Lu (2022).
8 Similar examples include the Puerta del Sol, initially a central landmark of the Spanish Indignados' anti-austerity movement, which was later renamed as Vodafone Sol" in a €3 million deal with the British mobile giant in 2013, rebranding the focal point of the anti-austerity protests with the name of a global phone company scrubbing its history out with global capitalism.

References

Alboher, M. (2007). *One person/multiple carers: A new model for work/life success.* New York: Warner Business Book.

Bennett, W. L., & Segerberg, A. (2013). *The logic of connective action: Digital media and the personalisation of contentious politics.* New York: Cambridge University Press.

Brenner, N. (2020). Is "tactical urbanism" an alternative to neoliberal urbanism? In: C. Courage et al. (eds), *The Routledge handbooks of placemaking* (pp. 312–321). New York and London: Routledge, Taylor & Francis.

Chuang (2022, September 12). *East lake for everyone: Records of a protest movement in Wuhan, 2010-2014.* https://chuangcn.org/2022/09/east-lake-for-everyone/

Cowan, G. (2004). Street protest architecture-dissent space in Australia. *Bad Subjects, 65*, 1–6.

Davidge, T. (2019, March 3). *How we campaigned in 2018.* Our City Our Square. https://bugle-ocelot-gpzg.squarespace.com/news-1/2019/3/3/how-we-campaigned-in-2018

Duff, C. (2017). The affective right to the city. *Transactions of the Institute of British Geographers, 42*(4), 516–529.

Elwood, S., & Hawkins, H. (2017). Intradisciplinarity and visual politics. *Annals of the American Association of Geographers, 107*(1), 4–13.

Everyone's East Lake. (2010, n.d.). http://www.donghu2010.org/2010/

FahadThetics (2017, October 10). https://www.youtube.com/watch?v=kQcq3rpne78

Frith, J., & Richter, J. (2021). Building participatory counternarratives: Pedagogical interventions through digital placemaking. *Convergence, 27*(3), 696–710.

Groys, B. (2014). On art activism. *e-flux Journal, 56*, 1–14.

Harvey, D. (2008) The right to the city. *New Left Review*, 53, 23–40.

Hou, J. (2010). *Insurgent public space: Guerrilla urbanism and the remaking of contemporary cities.* London and New York: Routledge.

Jacobs, J. (1961). *The death and life of great American cities.* New York: Vintage Books.

Latour, B. (2004). Why has critique run out of steam? From matters of fact to matters of concern. *Critical Inquiry, 30*(2), 225–248.

Lesh, J. (2021). Melbourne's Federation Square and its Heritage Discontents, 1994–2002. *Fabrications, 31*(1), 109–138.

Li, J. (2022, May 14). *Two Dragon King temples and the disappearing East Lake.* Tangent Essays. https://www.tangentessays.com/essays/two-dragon-king-temples-and-the-disappearing-east-lake-part-1/

Loopmans, M., Cowell, G., & Oosterlynck, S. (2012). Photography, public pedagogy and the politics of place-making in post-industrial areas. *Social & Cultural Geography, 13*(7), 699–718.

Mai, D. (2014). Public ruins. In Hui, Y (ed) *Creative space: Art and spatial resistance in East Asia.* Taipei: Yuanzhuo Elite.

Mattingly, D. C. (2019). *The art of political control in China.* Cambridge, UK: Cambridge University Press.

Novacevski, M. (2017, December 22). *Like it or loathe it, here's why Apple doesn't need a planning permit for its Fed Square store.* The Conversation. https://theconversation.com/like-it-or-loathe-it-heres-why-apple-doesnt-need-a-planning-permit-for-its-fed-square-store-89527

Page, T. (2020). *Placemaking. A new materialist theory of pedagogy.* Edinburgh University Press.

Papacharissi, Z. (2015). *Affective publics: Sentiment, technology, and politics.* New York: Oxford University Press.

Pavlićević, D. (2019). *Public participation and state building in China: Case studies from Zhejiang.* London and New York: Routledge.

Postill, J. (2018). *The rise of nerd* politics: *Digital activism and political change*. London: Pluto Press (Anthropology, culture & society).
Poxiansheng. (2012, n.d.). *An Interview with Maidian and Zijie*, Poxiansheng. http://donghu2010.org/?p=154
Puente, S. N. (2011). Feminist cyberactivism: Violence against women, internet politics, and Spanish feminist praxis online. *Continuum: Journal of Media & Cultural Studies*, *25*(3), 333–346.
Rancière, J. (2003). Politics and aesthetics an interview. *Angelaki*, *8*(2), 191–211.
Shin, H. B. (2013). The right to the city and critical reflections on China's property rights activism. *Antipode*, *45*(5), 1167–1189.
Stephens, D. (2017). *Reengineering retail: The future of selling in a post-digital world*. Vancouver: Figure One Publishing.
Stewart, J., & Lithgow, S. (2015). Problems and prospects in community engagement in urban planning and decision-making: Three case studies from the Australian Capital Territory. *Policy Studies*, *36*(1), 18–34.
Stiegler, B. (2018). *The Neganthropocene*. London: Open Humanities Press.
Suarez-Villa, L. (2012). *Technocapitalism: A critical perspective on technological innovation and corporatism*. Philadelphia: Temple University Press.
Suarez-Villa, L. (2021) *Globalisation and technocapitalism: The political economy of corporate power and technological domination*. London: Ashgate.
Yao, H. (2010, March 25). An Investigation into OCT's Development of East Lake in Wuhan: 450 *Mu* [27 hectares] to Be Filled, Fears of Environmental Degradation. *Time Weekly*, 25 March 2010 [Mandarin].
Xiao, J., & Lu, I. F. (2022). Art as intervention: Protests on urban transformation in China and Australia. *Journal of Urban Affairs*, *44*(4–5), 504–523.
Zuboff, S. (2019). *The age of surveillance capitalism: The fight for a human future at the new frontier of power*. London: Profile Books.
Zumthor, P. (2008). *Atmospheres*. Basel: Birkhäuser.

4 Organisational Assemblage

4.1 Introduction: Urban Economy, Care and Maintenance

Comprehensive examinations of urban political economy often recognise the intricate dynamics and transformation of urban economic development. Through historical analyses of American cities, Kantor (1987) elucidated three dimensions of the liberal-democratic urban political economy: urban economic dependency, popular control and intergovernmental development. These dimensions interacted variably over time, shaped by distinct urban development strategies. In recent years, scholars have increasingly scrutinised the impact of digital economies on urban development, positing that information and communication technologies (ICTs) function both as decentralised mechanisms to mitigate urban-regional disparities and as forces that accentuate centralisation (e.g., Carter, 2013; De Falco, 2019). Zhu and Chen (2022), in their study of Hangzhou, China, discovered that the digital economy dismantles the traditional core–periphery urban structure. Coupled with progressive urban development and economic policies, the influence of the digital economy on urban development surpasses conventional urbanisation processes, reinforcing the notion that governmental, market and individual actors have become pivotal in advancing urbanisation. The current literature has inadequately assessed the significance of two crucial aspects of the urban economy in contemporary digital cities.

First, while considerable emphasis has been placed on the economic impacts of infrastructural investments and 'specialist' urban activities, the persistent yet quotidian tasks of urban care and maintenance remain largely overlooked. Cities have been frequently depicted as growth machines (Logan & Molotch, 2007), driven by coalitions of developers, construction firms, financiers and real estate entities that champion and profit from urban development. David Harvey (1985) conceptualised the cultural economy as a tertiary circuit of capital, emerging from the overaccumulation crisis within the primary circuit of capital circulation, subsequently redirecting investments from secondary physical infrastructures to scientific, technological and cultural innovations. Similarly, creative placemaking, urban festivals and cultural events have been acclaimed as methods of urban renewal by remediating

DOI: 10.4324/9781003407973-4

a range of disorderly urban conditions (Cudny, 2016; Di Vita & Morandi, 2017; Redaelli, 2018). Nonetheless, the ongoing labour of urban repair, which discursively produces place through constant labour and mitigates unwelcome changes, can appear 'unnoticeable' or 'trivial' (Benson & Jackson, 2013). This includes tasks such as infrastructural maintenance, events programming and financial operations. The marginalisation of these activities may stem from the biased perception of care and maintenance as financial loss rather than profit (Considine, 2022), reflecting a hierarchical, prejudiced and often gendered understanding (Mattern, 2018).

Second, the transformative potentials of networked individuals within the digital urban economy have been vastly underestimated. Yochai Benkler (2006) posits that the Networked Information Economy (NIE) signifies a transformative shift from the industrial information economy by enabling decentralised and coordinated actions through 'distributed, nonmarket mechanisms that do not depend on proprietary strategies' (p. 3). The NIE opens a 'technological-economic feasibility space', amplifying the impact of peer-based commons and decentralised individual actions in the socio-economic production processes. Shoshana Zuboff (2019, p. 129) describes the 'extraction imperative' of surveillance capitalism, wherein human experience is mined as raw material and transformed into proprietary behaviour surplus, crafted into predictive products that fuel the burgeoning behavioural futures market aimed at sustaining dominance and maximising profit. The extraction imperative not only applies to the digital realm but also permeates the urban ecosystems where ubiquitous media and surveillance architecture predict patterns, create business cases and justify decision making in the processes of digital placemaking. Networked and platform urbanism necessitates that marketers and administrators heed popular sentiment, feedback and value propositions. The Chinese phenomenon of *wanghong* urbanism exemplifies the cyclical interaction between digital social media and the urban built environment, where a place or spatial practice gains online popularity, spurs offline engagement and is subsequently replicated in new locations as a packaged experiential product (Zhang et al., 2022). Increasingly, wanghong urbanism actively participates in co-creating feedback loops (Bruns, 2008) to improve urban administration instantaneously, as illustrated by the barbecue city Zibo, which, after achieving media fame, has immediately developed public transportation and accommodations for tourists (Wang, 2023). As these new practices integrate into formal urban institutions and form stable social alignments, their roles warrant greater attention.

Aiming to maintain place vibrance, the organisational assemblage creates spaces for the exchange of labour and value. Care and maintenance are essential to resist undesired changes in an entropic material world, while creation and innovation aim for connection and growth in an interconnected environment. Effective place management increasingly depends on both voluntary and involuntary citizen involvement in the design, management, programming, evaluation and value co-creation (Osborne et al., 2016). Thus, the organisational assemblage interlinks with market-driven agents, not entirely replicating the techno-capitalism agenda. For Amin and Thrift (2016), urban prosperity arises from the competitive advantage of 'spatial concentration of people, firms and institutions constituting the supply and demand base for growth, efficiency, and innovation' (p. 101). Unpacking the urban economy requires understanding how these agglomerations are influenced by multiple networks, flows, agencies and material provisions.

In this chapter, I analyse Federation Square's organisational assemblage, focusing on the Hack Fed Square event, its problem identification, collaborative networks, the rationales underpinning its organisation and the evaluation of the activity. I also discuss the wanghong city in China, highlighting the necessity of recognising placemaking labour as integral to the organisational assemblage. In particular, public participation is intrinsically woven into the fabric of digital city programming. Establishing a feedback loop for iterative design enhancement serves not merely to mitigate public discontent and uphold public relations. it constitutes a core element of the programs and activities delineating contemporary urban public spaces. This dynamic engenders a sense of co-production in urban environments, offering an opportunity to reconfigure public services. This chapter argues that organisational assemblages constitute spaces of potential, intricately woven with organisational constraints, multi-scalar complexities and negotiated compromises.

4.2 Hack Fed Square: An Event Overview

Hackathons have evolved beyond their software engineering roots, typically described as intense collaborative events where predefined problems are solved within a limited timeframe (Zapico et al., 2013). The term 'hackathon' itself reflects this shift, combining 'hack' with 'marathon' to signify enthusiastic collaboration rather than cybercrime (Nissenbaum, 2004; Wark, 2006). In recent years, 'hacking' has come to represent open and innovative thinking applied to various fields beyond computer sciences (Maalsen & Perng, 2016). Urban hackathons are a derivative form of civic hackathons that adopt collaboration and experimentation in solving urban development and renewal problems (Pogačar & Žižek, 2016).

Hack Fed Square was an urban design hackathon initiated by Federation Square's management organisation, Fed Square Proprietary Limited (FSPL), and facilitated by Melbourne-based marketing firm Marketing Entourage. The hack also collaborated with and featured in Melbourne Knowledge Week (MKW), which is the City of Melbourne's (COM's) annual festival of ideas for a smart and innovative city. According to Hack Fed Square's event page:

> Hack Fed Square, is a design hack that brings together passionate thinkers, makers and creators to shape the future of our iconic public square ... One of the key questions the hack hopes to address is: *How*

> *can we ensure Fed Square remains one of Melbourne's key cultural landmarks and iconic meeting places – a space where creativity, curiosity and inspiration collide?*
>
> (Fed Square, 2018)

The event invited 60 participants to form teams, attend an information session and participate in an eight-hour 'mental sprint.' Teams would then present their solutions to a panel of judges including figures from FSPL, the COM and Melbourne's start-up scene. The competition was open to the public and livestreamed through various channels.

Three preliminary observations emerge from initially examining this event. First, Hack Fed Square was led by Federation Square's management organisation, FSPL, rather than the general public. This contrasts with early hackathons, which often emerged from outside of formal and established structures. Early accounts of urban hackathons referred to distinct and 'grassroots' initiatives without government or corporate involvement, such as urban exploration, a subculture that sought hidden or abandoned urban spaces (e.g., Garrett, 2013). Hack Fed Square was not organised or paid for by the public. Maalsen (2021, p. 2) has observed that the 'rise of the smart city, platform urbanism, and the innovation economy has repositioned the hack, not solely as a site of citizen pushback but a tool of governance, entrepreneurialism and civics'. In line with this observation, the remit of Hack Fed Square was embedded in a larger urban environment sustained under the image of Melbourne as an innovative and creative city. Hack Fed Square has become a celebratory proof of that urban image rather than a challenge to the status quo (e.g., Friesinger et al., 2010).

Second, Hack Fed Square required significant labour and resources in planning, organising and hosting the event. Organising a hackathon involves various logistical aspects of event management (Zapico et al., 2013). For Hack Fed Square, this included collaborating with partners like Marketing Entourage and MKW to develop an event plan and program, sending callouts and invitations (to the public and the panellists), liaising with the panellists and stakeholders, preparing for background material and booklets, booking all-day catering and venues, providing appropriate facilities (for presenting information, livestreaming, participant prototyping, all-day video filming, interviewing and internet connection), publicising the event and potentially seeking sponsorship. Despite being a relatively smaller-scale event within the 2,560 events planned, hosted or facilitated by FSPL in 2019, Hack Fed Square required and mobilised substantial resources, planning and workforce from an authorising environment.

Third, the hackathon's focus lacked clarity. Hackathons typically target predefined questions with clear deliverables, such as prototypes or a piece of code. However, the question of maintaining Federation Square's significance as a cultural landmark and meeting place offered an ambiguous prompt.

The event description did not specify the desired outcome: would it be an app, an exhibition or simply co-designed ideas? Further, it remained unclear if and how winning pitches would be implemented in reality, potentially limiting participant engagement and the overall impact of the event.

4.3 Why Hack Fed Square?

The stated goal of Hack Fed Square, held in May 2019, was to co-design the public square's future. However, a closer examination reveals deeper motivations. Here, we explore three possibilities: a response to Apple's withdrawal, a crisis public relations exercise and a disguised cry for help.

4.3.1 A Closure for Apple

Hack Fed Square seemingly acknowledged the closure of the Apple town square plan by inviting the publics to co-design alternative futures for Federation Square. FSPL's initial rationale, detailed in a since-deleted media release (Fed Square, 2017), celebrated Apple's selection of Melbourne to establish the 'Global Flagship Store', praising it as more than just 'any regular Apple shop' (Fed Square, 2017) but the only one of the six flagship stores in the southern hemisphere. Former FSPL CEO Jonathan Tribe (2018) lauded Apple in an opinion piece, claiming it would be not only a new tenant but also 'a new age arts and technology hub' with digital technologies that 'connects and includes people without prejudice' and 'a great complement to the square's civic and cultural fabric'.

Apple's proposed contribution included a weekly program of free educational events in technology, arts and innovation, conducted through their 'Today at Apple' program, led by trained staff and creative workers like artists, photographers and musicians. FSPL and Apple even initiated discussions in 2018 to identify projects of mutual benefit, leveraging emerging technologies (Annual Report 2017–2018, 2018). Most FSPL employees, despite limited involvement in negotiations, expressed excitement about Apple's potential. For example, Deborah Allanson, program director for the Digital Façade, commented on the Apple brand's association quality, innovation, culture and a younger mindset (Allanson, Interview, 28 May 2018).

Despite minimal involvement and decision-making in having Apple as a new commercial tenant, FSPL swiftly incorporated the Apple town square plan into its strategic imperatives. One of five strategic imperatives outlined in the 2018–2021 Corporate Plan (Annual Report 2018–2019, 2019, p. 17) focused on preparing Federation Square to 'open up to the Yarra River and accommodate new and innovative offerings'. The primary action point was integrating the Apple town square into the precinct. Originally, this change aimed to improve Yarra River access. Several accommodation configurations were already under way, including:

> Consolidation of Special Broadcasting Service (SBS) from two levels of the Alfred Deakin Building into a single refurbished level.
>
> Construction of a new staircase to Level 2 of the Alfred Deakin Building to provide greater external connectivity, access and utilisation.
>
> Refurbishment of Chocolate Buddha in the Alfred Deakin Building, and development of the Pop-Up Buddha restaurant in the Yarra Building.
>
> (Annual Report 2018–2019, 2019, p. 36)

However, Federation Square's nomination for heritage protection led Heritage Victoria to impose an Interim Protection Order, halting development. This prompted FSPL to apply for a Yarra Building demolition permit, which was ultimately denied in April 2019. With this decision, the Apple project became unfeasible, rendering FSPL's significant effort null and forcing a strategic re-evaluation and re-engagement with the publics.

4.3.2 A Crisis Public Relations Response

Following Apple's withdrawal, FSPL faced public discontent. Hack Fed Square could be interpreted as a response aimed at restoring Federation Square's reputation after a year of public unrest. The event emphasised an image of Federation Square as an open, fun and community-engaged space. Jonathon Tribe, FSPL CEO (2016–2019), stated:

> Hack Fed Square is a fantastic initiative that continues Fed Square's history of seeking out and responding to the input and ideas of members of our community in how to ensure that we remain Melbourne's premier civic space. This event is a truly unique opportunity for members of the public to provide ideas that will shape the future of Fed Square and help us maintain our position as an iconic meeting place and cultural landmark.
>
> (Fed Square, 2018)

Hackathons prize the logic of playfulness, creativity, efficiency and optimisation (Maalsen, 2021). They are often praised as social events gathering diverse groups and participants to build inclusive communities (Zapico et al., 2013). Tribe's remarks echoed similar sentiments, painting Hack Fed Square as a user-centric event and reinforcing a community-focused image of Federation Square. This focus on generating a positive image could be seen as an attempt to mitigate the negative impacts of the Apple controversy, which lacked community approval and standard public consultation processes. Anthony Townsend (2013) argues that pop-up, open-source smart cities rely on a combination of creative talents, receptive audiences of civic hacker culture and authorising environments and rules to promote innovative infrastructure. Hack Fed Square served the branding function as an event to restore Federation Square's image and reputation.

4.3.3 A Cry for Help

Hack Fed Square can also be understood as a subtle cry for help from FSPL, revealing its underlying problem. Those familiar with Federation Square's pre-hackathon predicament recognised that ensuring its status as a cultural landmark and iconic meeting place posed a potential threat to losing that very status. Interviews with past and present FSPL managers suggested three critical issues threatening Federation Square's sustainability, all of which contributed to FSPL's acceptance of the Apple proposal.

First, FSPL's business model jeopardised Federation Square's operational sustainability. An atypical message from the Chair and CEO in FSPL's 2018–2019 Annual Report (2019, p. 6) alerted the public to Federation Square's problems:

> While Victorians, Australians and international visitors know Fed Square, they may not fully appreciate the unique challenges of the precinct. Fed Square is an independent, commercial company, answerable to the Minister for Tourism, Sport and Major Events with an expectation that its public role is financed through commercial revenue and the leases of the institutions, not-for-profit organisations and businesses that call Federation Square home. With limited avenues for generating revenue, significant community expectations for protection, access and affordability and increasing costs associated with maintenance of Fed Square's innovative infrastructure, Fed Square's operating model requires a successful balancing of these important, but at times uncommercial, objectives.

FSPL relied on revenue from tenancy, car parking, commercial events, sponsorship and grants to offset its operational and personnel costs. Despite shifting placemaking strategies to incorporate more events and programs, this revenue structure remained unchanged since its inception, leading to a growing deficit. Additionally, FSPL could not raise the rent of its major cultural tenants like National Gallery of Victoria (Australia) and Australian Centre of Moving Images due to their subsidised, non-commercial rates as Victorian cultural institutions.

Second, the rising maintenance costs due to infrastructural deterioration were not budgeted for by either the Victorian Government or FSPL. William Blakeney, FSPL general manager for corporate services, explained that the 15-year-old infrastructure, like leaking pipes and increasing fire risks, has strained FSPL's financial reserves. Millions of dollars have been spent, but the problem remained; hence, the focus was on wanting to be more commercial. (Blakeney, Interview, 28 May 2018).

Matthew Jones, former FSPL general manager for digital futures and event operations, shed light on the Victorian Government's reluctance to address FSPL's exponentially growing loss projection and financial needs. As FSPL faced significant unanticipated expenditures that would have a long-term impact on their budget, the Victorian Government prioritised funding based on public need. However, long-term infrastructure upgrades can be challenging to fit into annual budgets and negatively impact the political optics. As the Government is committed to Federation Square's success, it would be more likely to address these issues through a significant investment in forming a masterplan or a revised funding model when the problem gets severe in the future.

Last, strategic collaboration between FSPL and its existing cultural tenants proved challenging. Former FSPL manager Jones highlighted the reluctance of traditional arts institutions to operate with shared budgets and programs, hindering collaboration (Jones, Interview, 30 April 2018). Faced with financial strain, infrastructural deterioration and the difficulty of collaborating with existing tenants, FSPL began to find the Apple plan increasingly attractive. FSPL manager Blakeney commented that FSPL was 'hoping for some leadership or partnership with them [Apple] in terms of their resources, huge, compared to anything [else on offer]' (Blakeney, Interview, 28 May 2018).

With Apple's potential market-rate rents and resources, including the promised Yarra Building redevelopment, no longer on the table, Hack Fed Square can be seen as an attempt by FSPL to raise public awareness about its funding problems.

4.4 Participation, Observation and Reflection

This section explores my experience at Hack Fed Square. I delve into the hackathon processes, the pitch contest result and my reflections.

4.4.1 *Information Night and 'Anything but Square'*

A Google alert in May 2019 led me to discover Hack Fed Square, a two-day co-design event inviting the public to participate in shaping Federation Square's future. After registering and answering questions about my occupation, reasons for attendance, expectations for Hack Fed Square and the frequency and whereabouts of my visits to Federation Square, I was selected as one of 60 participants and was asked to commit to attending both the information briefing night and the design pitch competition.

The information night served as a crucial introduction. We encountered a combined team of more than 10 people representing the MKW programming team, FSPL's events and marketing teams, judging panel representatives and Marketing Entourage facilitators. The challenges facing Federation Square were laid bare. Suzana Bishop, FSPL's General Manager, highlighted increasing competition from other precincts, limited funding and annual losses of

A$6 million. She emphasised the need to explore cultural expressions beyond traditional arts, potentially encompassing food, wine, dance and leisure activities.

A surprise awaited us with Amy de Groot, FSPL's Head of Marketing, unveiling 'Anything but Square', the rebranding campaign based on a 2019 visitor survey conducted by FSPL's marketing team. This 30-question online survey, the largest FSPL had ever undertaken, gathered more than 1,300 responses from Melbourne residents. The results, as de Groot explained, revealed a desire for visitors to gain a sense of ownership and experience a kaleidoscope of activities, including leisure, dining, retail and immersive technological experiences.

Trent Clews-de Castella, CEO of PHORIA, a Melbourne-based immersive tech start-up, gave a short presentation about how augmented reality (AR) and virtual reality could reshape the urban vision. The power of digital technologies could be harnessed to create a digital twin of the physical site, allowing the exploration of responsive and connected opportunities in a hybrid format.

Samantha Hurley, co-founder of Marketing Entourage, delved deeper into the Anything but Square rebranding campaign and introduced the participants to 'Freda Square' and 'Fred Square'. These two avatars represented the typical female and male visitors: middle-aged, married, white Victorians working full time, families earning over A$100,000 annually and spending $A20 to 50 dollars per visit. Hurley emphasised designing a customer-centred experience through a participatory, human-centred process, encouraging the participants to consider these customer profiles during brainstorming.

Following introductions to more FSPL employees from various divisions, we were encouraged to connect with people around us and form teams of six. Our diverse group comprised individuals from Australia, India, Germany, China, Colombia and Israel. Three of us pursued PhDs in architecture, transportation systems and digital placemaking. The rest were finance, urban planning and tourism professionals. Looking at Freda and Fred Square, we joked to each other, 'Well, we are not like them at all'. Looking around, I observed a similar diversity among participants, many in their 20s and 30s. Some appeared more senior, while a pre-formed group of bachelor design students stood out. The atmosphere felt competitive, with the prospect of a design pitch potentially resembling a throat-cutting competition among such highly educated professionals.

Fuelled by an abundance of cheese and nibbles, a team spirit emerged. Deciding on 'Square Cheese' as our name, we acknowledged the irony—cheeses are generally round, not square, reflecting our unique and diverse nature as the 'odd bunch'. After the information session, we embarked on a collective tour of Federation Square. Strolling past the dimly lit River Terrace, connecting the square to the Yarra River, the architects and planners among us voiced concerns. The severed connection between the square and the river disrupted the spatial flow. We decided to explore this as a potential focus for our brainstorming session the following day.

4.4.2 The Hacking Day

Reuniting at Federation Square's Deakin Edge, team Square Cheese joined others in our collaborative brainstorming and designing processes. A booklet offered all participants further information about the objectives and processes of Hack Fed Square.

The objectives of Hack Fed Square were reiterated: positioning Federation Square as a world-class, iconic destination appealing to visitors and partners through an intriguing voice and deploying unique, immersive experiences for long-term sustainability (Hack Fed Square Material, 2019). The emphasis on visitors, partners and long-term sustainability echoed the Civic and Cultural Charter's objectives to provide visitors with an appealing experience. However, Hack Fed Square seemed more economically ambitious, prioritising these factors over civic and cultural cohesion, a focus of FSPL's Cultural and Civic Charter (Charter).

The Anything but Square campaign further specified three target audience groups: 'Hopefuls' (young, professional parents seeking cultural experiences for relaxation and de-stressing), 'Conscious Capitalists' (free-spirited hedonists seeking community experiences and intellectual stimulation) and 'Cool-Hunters' (young originals craving unique, money-can't-buy experiences). Each group supposedly represented distinct needs and desires. The campaign's envisaged goal is to bring the 'Hopefuls', invite the 'Conscious Capitalists' back to Fed Square and get the 'Cool-Hunters' to trade up and spend more. As will be reflected on later in this chapter, the narrowness of the desired visitors—young, wealthy, cool-hunting people, able-bodied—captured a small fraction of the population, and the Indigenous community, seniors, children, small multicultural groups and the lower-spending population were excluded from the design parameter. The judging rubric, unveiled that morning, assessed project ideas based on four criteria: financial sustainability, civic and cultural value, brand alignment (adherence to the Anything but Square campaign) and visitor engagement. Evidently, finance and brand making combined outweighed the Charter's emphasis on social cohesion.

Square Cheese spent the next eight hours among coffee cups, markers, fluorescent Post-it notes, Lego, phones, sticks and other gadgets. We collectively and unconsciously decided to skip the step of understanding the target audience with a self-righteous belief that 'we should be Federation Square's target audience'. Initially, the disconnection between Federation Square and the Yarra River continued to hold our attention. One team member proposed a series of public slides extending from the Yarra Building towards the river, creating a playful and interactive pathway. Others suggested collaborating with Airbnb and organising open-air camp nights on Federation Square's paddock or in the underground tunnels. While these ideas resonated with us, concerns arose regarding feasibility and cost-effectiveness.

Moving beyond the river connection, the team considered utilising the iconic Digital Façade. One creative idea involved generating mist on the

Yarra River and using it as a projection screen and connecting it with the Digital Façade. People would have to walk back and forth, using their bodies to connect the square and the river. Another proposal envisioned a series of light-and-sound installations along the façade, responding to visitor movement and the surrounding environment. However, like many other teams' ideas, ours relied heavily on technology and interactivity. The provided materials—markers, sticky notes and basic prototyping tools—proved insufficient for effectively presenting such complex concepts.

The pitch contest took place at the stage area of the Deakin Edge and was livestreamed through the online channels of Federation Square and MKW. All team members were invited onto the stage. After the pitch, the judging panel would ask one or two questions. Our pitch, themed around 'home' for everyone, was a nervous improvisation based on nothing that Square Cheese had carefully rehearsed. After all the teams had presented their ideas, the judges went away to discuss their choice of the winners. The audiences watching the livestream were invited to vote for their favourite for the 'People's Choice Award' through real-time polling.

The judges returned to announce the winners. Claiming no victory for Square Cheese did not influence our morale. Before calling it a day, we sincerely applauded the chosen winners who received cash rewards. Unfortunately, there was no follow-up information regarding the implementation of these ideas.

4.4.3 Researcher Reflections

My initial reflections on Hack Fed Square acknowledge the event as an assembling force to bring people together and mobilise different kinds of labour. The winning pitches extended FSPL's arts and cultural programming placemaking strategies into a hybrid and user-generated space. However, the market-driven logic of the hackathon cast some doubt.

First, Hack Fed Square broadened participation and attracted a diverse group of participants, fostering camaraderie and a sense of community. This aligns with how hackathons engage the public with data and information typically reserved for professionals (Zapico et al., 2013). FSPL generously shared its challenges, value proposition and market research insights. While urban hackathons often attract people with more socio-cultural capital (Maalsen & Perng, 2016), the demographic was more heterogenous than expected. This diversity led to thoughts around designing for difference rather than designing for excellence, exemplified by Square Cheese considering playful experiences for people from diverse cultural backgrounds. These ideas stemmed from a naïve yet sincere sense of ownership and empathy for diverse user groups, potentially explaining why we did not win the pitch contest—we may not have addressed the target audience (the marketing personas, the 'Hopefuls', 'Conscious Capitalists' and 'Cool Hunters') effectively. However, the exercise was valuable, promoting interdisciplinary learning and relationship building.

Square Cheese remained connected via WhatsApp for months after the event. Hack Fed Square provided a chance for community formation and place attachment, both key components of placemaking initiatives.

Second, Hack Fed Square successfully mobilised multi-scalar labour. Although the hackathon was not a large-scale event, it brought together the extensive labour and collaboration of governments, companies and organisations, such as the COM, FSPL, Marketing Entourage, Phoria and Uvideo (Marketing Entourage's filming partners). The marketing, events and public engagement teams of FSPL were especially engaged, alongside the technical, security, cleaning and catering staff that tend to be ignored. It also successively mobilised 60 participants to bring their own interest, professional knowledge, creativity, worldview, passion and understanding of Federation Square and Melbourne to the hackathon. The teams actively engaged with FSPL's challenges and demonstrated a genuine desire to contribute to solutions. This mobilisation of multi-scalar labour highlighted the intricate web of stakeholders involved in shaping public spaces.

Third, Hack Fed Square extended FSPL's creative and digital placemaking initiatives with emerging technologies. Since 2005, FSPL has spearheaded the use of the Big Screen (later the Digital Façade) to engage diverse communities through digital storytelling. Rather than merely broadcasting content from major media organisations[1] or used for advertising, the Big Screen partakes in 'civic partnership' (McQuire, 2017, p. 139) by sourcing user-generated content and constantly experimented ways of audience engagement and place activation through newly afforded technologies. Throughout the years, the Big Screen experimented with contextual messaging, interactive dance battles, multiplayer and multimedia games, and now exploring immersive technologies. This showed executive acceptance of how individuals and communities could enrich the cultural programming of Federation Square. The winning pitchers highlighted the value of art, culture and Melbourne's unique history and its Indigenous heritage. Maalsen (2021) recognised that one significance of urban hackathons was not just to utilise various technologies but to translate the computational logics to the urban. In this line, Hack Fed Square continued to honour the user-generated spirit and collective intelligence underpinning FSPL's cultural programming. Hack Fed Square cultivated playful and exploratory urban experiences instead of technocratic and solutionist forms of placemaking (de Waal et al., 2020).

Despite its strengths, Hack Fed Square's process was not without limitations. The dominant market orientation introduced a layer of complexity. Participants were provided with existing customers, Freda and Fred Square, and tasked with identifying additional target audiences. This approach potentially limited the inclusivity of design solutions for a public space. While such strategies might work for niche markets and specialised products, a public square serves a broader purpose. Enlarging target audience engagement through audience segmentation could deviate from the public interest and exclude certain multicultural communities prone to marginalisation.

This raises concerns about the potential tension between economic considerations and the core function of public spaces to foster social cohesion and cater to a diverse citizenry.

One reason for the market orientation seen in the hackathon is the facilitator, Marketing Entourage, a digital marketing company. Traditional co-design practitioners, such as urban designers, tend to face a difficult position: taking an uncritical stance can lead to normalising the dominant logic of the elite, while taking a clear critical stance can result in losing access and trust from stakeholders and ultimately result in being sidelined from the industry (Kaethler et al., 2017). The choice was less difficult for Marketing Entourage, which was hired to help FSPL increase site visitation rather than contemplating the dominant logic of the elite.

Another explanation could be FSPL's budgetary constraints and a genuine need to boost revenue and visitation. Public infrastructure projects often have hidden costs associated with ongoing maintenance and programming. Hack Fed Square might have been an attempt to address these concerns through a market-driven approach. However, it is crucial to acknowledge that the delivery of public benefits for a space extends beyond construction. The financial and political realities embedded in public infrastructure projects create ongoing challenges for place managers responsible for maintenance, programming, and fostering the community that inhabits the space. Acknowledging the placemaking labour and financial needs of FSPL was crucial to understanding placemaking as delivering public benefit in a continuum.

4.5 The Embedded Complexities of Evaluating Hack Fed Square

Official evaluations of Hack Fed Square were brief. FSPL's annual report called it a success for community involvement and creative potential (Annual Report 2018–2019, p. 24). Marketing Entourage highlighted the collaborative nature of the event (Marketing Entourage, 2019). Indeed, as an empirical event, the two-day event went smoothly without technological interruption or security hazards, with a full house of participants and panellists. However, after scrutinising Hack Fed Square under multiple frames, we would find that the evaluation of such an event can be more complicated.

4.5.1 The Ambiguity of Hackathons

First, urban hackathons like Hack Fed Square face inherent ambiguities. Invoking Agamben (2008), Maalsen (2021, p. 8) argues an urban hackathon can function as 'a permanent state of exception' that generalises the condition to address the rigidities and inflexibility of urban systems. Urban hackathons can provide a solicited response to prompted urban challenges, but they lack the power to make significant structural changes because they are designed to work around predefined parameters and function within the constraints of the existing systems.

My observation differed from some criticisms of hackathons, like gender and class imbalance (Maalsen & Perng, 2016) or under-representation of the population (Mattern, 2016). On the contrary, I observed a mismatch between the diversified participants and Federation Square's target audience groups, which were defined by a rather monocultural and comic union of Freda and Fred Square. This observation aligned with Perng's (2019, p. 421) comment that hackathons tend to produce 'a neoliberal citizenship in which citizen efforts are constrained by the broader privatisation of the city and tinkering for social purposes is limited'. No team seemed to question the market orientations stipulated by the remits of Hack Fed Square. Nobody was seemingly bothered or confused by Freda and Fred Square.[2] More importantly, in the rapid procession of the event, nobody paused to think whether market-driven design and persona-based marketing aligned with the public space ethos enshrined in Federation Square's Civic and Cultural Charter.

Second, Hack Fed Square was embedded in the context of increasing urban innovation and entrepreneurialism, blurring the lines between co-design and market forces. While the event explored market-driven solutions to improve place vibrancy and financial sustainability, this focus was also heavily influenced by FSPL's existing, unsustainable business model and long-standing financial problems, which existed well before the Hack Fed Square event. While the overt market logic and obscure outcomes of the hackathon can be critiqued, it is necessary to understand the context of Fed Square's ongoing challenges, particularly in light of the abandoned Apple plan and the uncertainties surrounding the Federation Square Review. At that time, Federation Square remained a self-funded commercial entity with an increasing deficit. This context helps explain why good ideas generated at the hackathon may face limitations during implementation due to FSPL's various restraints.

Third, Hack Fed Square was inevitably part of an urban world replete with increasing drives and appetite for innovation and entrepreneurialism that both emerge from and reinforce neoliberal urbanism. This situation added to the ambiguity of Hack Fed Square and manifested as a delicate relationship between urban co-design hackathons and the market. Early research on co-design events and urban hackathons has similar narratives. Co-design, or participatory design, was rooted in 'an activist tradition aspiring to increase democratic participation of diverse societal groups in design activities related to public space, services, systems or policy' dating back to the 1960s (Huybrechts et al., 2017, p. 145). Similarly, urban hackathons were positioned as oppositional strategies and radical interventions to the privatisation of public spaces (Friesinger et al., 2010). This narrative viewed democratic social change and market engagement as incompatible pursuits (Papanek & Fuller, 1972). Since then, a broader shift has taken place as many traditional aspects of the public domain have shifted to the private domain (e.g., Graham & Marvin, 1995; Harvey, 1990). Such a shift means co-designs no longer occur across previously contrasting public and private spheres but in a blurred zone where values of the common good, democracy, market and entrepreneurialism increasingly intertwine. Rather than hastily declare that design hacks are democratising deeds or entirely compromised by market forces, we need to understand that Hack Fed Square was embedded in various urban complexities and compromises, reflecting the current state of co-design practices.

4.5.2 The Neglected Place Managers

Urban hackathons are often viewed and evaluated as participatory opportunities or neoliberal co-opts, but little attention is given to the crucial role of place management and the continuous placemaking labour it entails. While some scholars question whether the public participation in hackathons should be properly renumerated if it provides an informal path for procurement and problem-solving (Johnson & Robinson, 2014), the placemaking labour from citizens and communities has received much scholarly attention. For example, Sweeney et al. (2018, p. 3) highly regarded citizen-led placemaking initiatives assembling both material and human resources and 'practices of noticing, observing, feeling and communicating into making, maintaining and remaking places'. Conversely, place managers are often ignored or criticised for spearheading the neoliberal agenda by transforming the city into a marketplace (e.g., Ashworth & Voogd, 1990; Fincher et al., 2016; Shaw & Montana, 2016).

However, placemaking requires incessant labour of infrastructural maintenance and place programming under FSPL's remit. Similar to Anderson's (2012) assemblage thinking about the coastline being continually maintained and stabilised by waves, surfers and the action of surfing, Federation Square requires ongoing labour to sustain its physical and digital assets, program events, ensure security and keep it clean. This labour is a cyclical, repeated process that carries a paradoxical sense of futility and necessity (Arendt, 1958). It is futile because it is repeated work against the entropic tendency of natural decline. Federation Square's infrastructure deteriorates whether it is a budgeted item or not. However, labour is also necessary in that FSPL's placemaking labour not only prevents the site from deteriorating but also allows it to function as a public sphere. Administration, logistics, cleaning, event management and infrastructural maintenance—deemed by some as trivial and inconspicuous activities—sustain a sense of place.

Throughout this chapter, I have emphasised the extensive and multi-scalar labour assembled by the Hack Fed Square event. Hack Fed Square functioned as a conduit for community formation, place attachment, encounters and experience on a personal level. It also highlighted a fraction of the ongoing placemaking labour required by FSPL. The event involved significant non-substitutable labour in 'assembling the necessary actors, materialities and agencies' (Sweeney et al., 2018, p. 7) internal and external to its organisation and subject to its various limitations. If these efforts are not considered, Hack Fed Square could be misjudged as a superficial co-optation or rebranding exercise. It would be dismissive if we only look at the rhetoric of Anything but Square without understanding that Federation Square is not free from empirical concerns and limitations (e.g., self-reliant operations and infrastructural deterioration). Similarly, we would not be doing justice to FSPL and the organisational assemblage of Federation Square (including FSPL and its collaborators) if we conflate a real financial concern related to the specific site with a broad-stroke frame of neoliberalism.

4.5.3 Doomed Public Participation?

Validating Arnstein's (1969) ladder of public participation, Cardullo and Kitchin's (2019) paper cast a gloomy outlook on smart citizen's urban participation. Curiously, Cardullo and Kitchin (2019) framed hackathons to promote and enact neoliberal citizenship despite varying degrees of 'citizen control'. They argue that citizen participation is often 'instrumental' rather than 'normative or political,' encouraging solutions within existing frameworks rather than challenging them (Cardullo & Kitchin, 2019, p. 18). While neoliberalism shapes urban landscapes as a conditional force that requires compliance (Brown, 2006; Dean, 2008, p. 48; Hesmondhalgh & Baker, 2013), I argue that neoliberal condition does not entirely cancel the significance of hackathons or terminate urban politics. If we understand power not as a stable position but as a relational state of affairs in a longer timeframe, we will not anticipate that hackathon participants could achieve 'citizen power' in a two-day event.

The organisational assemblage bears resemblance to what Paolo Cardullo (2020) terms the 'intelligent people cities' in which the capitalist cultural establishment of digital cities needs to be broken down into locally acceptable cultural assemblages maintained with ongoing care and collaboration. Acknowledging the agency of place management organisations in the organisational assemblage sheds light on their role as intermediaries between activists and governments. For example, FSPL could have addressed Federation Square's real predicament by proposing an honest problem in a less celebratory tone and grounding that problem within Federation Square's specific and multiple realities. Alternatively, the design problem of Hack Fed Square could have been defined by the community together with FSPL, which would offer a better chance for FSPL to harness collective urban intelligence. Either way would require treating structural problems not as defining conditions but as subjects of open discussion. The next chapter explores the Federation Square Review, a 'policy hack' with a longer timeframe, greater resources and tangible structural changes. In conclusion, echoing Maalsen (2021), I view hackathons as 'spaces for possibility' (Massey, 2005) inherently subject to complex compromises within a layered urban landscape.

4.6 Wanghong Cities

The phenomenon of wanghong (translated in English as internet celebrities) aligns with the expansion of the Chinese platform economy. The wanghongs are not merely content creators or platform entrepreneurs; they constitute an integrated dimension of the Chinese media platform ecology where creators cultivate communities by offering culturally or commercially viable products and services (Craig, Lin & Cunningham, 2021). As participatory short video platforms and e-commerce platforms such as Douyin, Kuaishou, Taobao and Red gained prominence, cities began to integrate branding strategies into their place narratives and placemaking efforts to drive online traffic and generate revenue. Wanghong cities are a hybrid of urban capital operations and current media consumption patterns.

In contrast to the traditional tourist hubs along China's eastern coast, western and inland cities such as Chongqing, Xi'an and Chengdu have emerged as the vanguard of wanghong (internet-famous) cities, formulating strategic partnerships with short video platforms for innovative city branding. In 2018, short videos of Chongqing on Douyin amassed a staggering 11.36 billion views, a virality that translated directly into tourism statistics. That year, Chongqing welcomed 597.24 million domestic and international visitors, generating a total tourism revenue of 434.4 billion yuan ($60.5 billion), marking year-on-year increases of 10.13% and 31.32% respectively (Qinghua University, 2018). Similarly, Xi'an, the ancient capital of the Tang Dynasty, has undergone a temporal transformation, redesigned as a Tang-themed cityscape to align with contemporary tourists' digital practices such as livestreaming, virtual flaneur and the 'clock-in' culture (Fan & Lin, 2022). Tourists navigate cultural and creative precincts reimagined to replicate ancient architecture, donning Hanfu (traditional Chinese clothing), greeting each other in traditional etiquette and immersing themselves in a carefully curated blend of nostalgia and excitement, entranced in a profound sense of pride and the geopolitical splendour of the Tang Dynasty.

Some scholars argue that the wanghong urbanism creates new forms of urban inauthenticity and social distinction and expedites gentrification, creating illusions of bourgeois urban aesthetics and romanticised 'Chinese Dream' (Ping, 2024). Other scholars observe that small- to medium-level cities, while being induced by state governments to adopt the attention-grabbing tactics, can utilise tactical authenticity on social media to create desires for consumption of identity performance, rather than succumbing to the 'Bilbao effect' (Banks, 2022). On the other hand, wanghong cities are believed to have an impact on the revitalisation of the local economy (Cunningham, Craig & Lv, 2019) by converting online traffic viewership into purchasing power through online commerce and advertising revenue (Han, 2021). Not only cities but rural regions and towns also compete to boost their economy by cultivating cultural intellectual property (IP) and integrating resources through technological innovation. Rongjiang Village Super

League, which started as an amateur football league in an obscure town in Guizhou, has grown into a significant hybrid sports carnival, attracting millions of online viewership and tourists and generating revenues for upstream and downstream industrial chains, such as hotels, restaurants, retailing and e-commerce. Under China's national development strategy of rural revitalisation (Chinese Government, 2018), major livestreaming companies, such as Douyin and Kuaishou, have been incentivised to conduct business with local governments and rural live streamers. Digital media technologies offer cost-effective ways for tourism destinations to marketise to more targeted customer segments supported by platforms' geo-tagging and big data analytics, which warrants disadvantaged regions and areas to participate in national and global competition (Law, Chan & Wang, 2019).

Wanghong cities also face their own financial problems on the balance sheet. For instance, the barbeque city of Zibo, which enjoyed an unexpected fame on social media platforms, experienced a sudden tourist boom. However, the urban ecology of Zibo, once China's important petrochemical and heavy industrial city, suffered from an outdated industrial structure and growth pattern dating back to the 1990s, with many power plants and old factories surrounding the train station in the city centre. Further, Zibo's local government has been in debt of 55 billion yuan ($7.8 billion), with a rapid annual increase of around 13 billion yuan ($1.8 billion), to which barbecue and other service industries could contribute minimally to the local fiscal revenue. The debt is circulated in the forms of municipal investment (chengtou) bonds created to raise capital for public projects or real estate in the municipalities through financing vehicles implicitly backed by local and central governments. As Chinese local cities' biggest revenue is from land leasing and the local fiscal revenue growth still relies on 'land finance', some local governments are less incentivised to transform to cultural or service-oriented cities. Rightly so, it would not be easy for other organisations or local governments to sustain the vibrance of wanghong cities outside its context, such as finance, infrastructure, degree and scale of industrialisation, labour market, consumption habits, culture and policy support.

4.7 Conclusion

In this chapter, I discussed the urban co-design hackathon Hack Fed Square organised by FSPL. With an in-depth ethnographic account and researcher's personal reflection, this chapter focused on how the event was being planned and organised, explored several reasons for its occurrence and reflected on the ambivalence and complexity of its evaluation. It found that the planning, organising and hosting of Hack Fed Square were embedded in a network of urban actors, materialities, agency and labour. Hack Fed Square could be seen as FSPL's seeking an alternative future without Apple as a tenant while repositioning Federation Square as a responsive, fun and open place for community engagement. It could also be seen as a revelation of FSPL's more profound problems that predate the hackathon itself, including financial crisis, infrastructural deterioration and unsustainable business model. On the one hand, the event manifested a strong commercial orientation through its problem framing, design parameters, logical approach, judging rubrics, timing and facilitators. On the other hand, the event translated the digital logic of user generation into urban programming and facilitated formulating of urban encounters and friendship communities and strengthening a sense of placeness and belonging in my participation. This chapter has portrayed organisational assemblages as a space of possibility embedded in its site specificities, environmental complexities and subject to its limitations and compromises. It further highlighted the importance of acknowledging the organisation assemblage and its ongoing placemaking labour as well as a need to evaluate urban hackathons and their political implications within the relevant organisational limitations and multi-scalar urban complexities.

Notes

1 For urban screens as public space broadcasting, see McQuire (2017), Chapter 4.
2 Notably, one of my interviewers, urban designer Jonathon Daly, has reportedly left Hack's Fed Square's information night in disgust due to these personas.

References

Agamben, G. (2008). *State of exception*. Chicago: University of Chicago Press.

Amin, A., & Thrift, N. (2016). *Seeing like a city*, Cambridge, UK: Polity Press.

Anderson, J. (2012). Relational places: The surfed wave as assemblage and convergence. *Environment and Planning D: Society and Space*, *30*(4), 570–587.

Annual Report 2017–2018. (2018). *Annual report 2017–2018*. Federation Square. https://fedsquare.com/about

Annual Report 2018–2019. (2019). *Annual Report 2018–2019*. Federation Square. https://fedsquare.com/about

Arendt, H. (1958). *The human condition*. Chicago: University of Chicago Press.

Arnstein, S. R. (1969). A ladder of citizen participation. *Journal of the American Institute of Planners*, *35*(4), 216–224.

Ashworth, G. J., & Voogd, H. (1990). *Selling the city: Marketing approaches in public sector urban planning*. London: Belhaven Press.

Banks, D. A. (2022). The attention economy of authentic cities: How cities behave like influencers. *European Planning Studies*, *30*(1), 195–209.

Benkler, Y. (2006). *The wealth of networks: How social production transforms markets and freedom*. New Haven and London: Yale University Press.

Benson, M., & Jackson, E. (2013). Place-making and place maintenance: Performativity, place and belonging among the middle classes. *Sociology*, *47*(4), 793–809.

Brown, W. (2006). American nightmare: Neoliberalism, neoconservatism, and de-democratisation. *Political Theory*, *34*(6), 690–714.

Bruns, A. (2008). *Blogs, Wikipedia, second life, and beyond: From production to produsage* (Vol. 418). Lausanne: Peter Lang.

Cardullo, P. (2020). *Citizens in the 'smart City': Participation, Co-production, Governance*. New York and London: Routledge.

Cardullo, P., & Kitchin, R. (2019). Being a 'citizen' in the smart city: Up and down the scaffold of smart citizen participation in Dublin, Ireland. *GeoJournal*, *84*(1), 1–13.

Carter, D. (2013). Urban regeneration, digital development strategies and the knowledge economy: Manchester case study. *Journal of the Knowledge Economy*, *4*, 169–189.

Chinese Government. (2018). *Strategic plans for rural revitalisation (2018–2022)*. https://www.gov.cn/gongbao/content/2018/content_5331958.htm

Considine, M. (2022). *The careless state: Reforming Australia's social services*. Melbourne: Melbourne University Press.

Craig, D., Lin, J., & Cunningham, S. (2021). *Wanghong as social media entertainment in China*. New York: Palgrave Macmillan.

Cudny, W. (2016). *Festivalisation of urban spaces: Factors, processes and effects*. Berlin: Springer.

Cunningham, S., Craig, D., & Lv, J. (2019). China's livestreaming industry: Platforms, politics, and precarity. *International Journal of Cultural Studies*, 22(6), 719–736.

De Falco, S. (2019). Digital and urban spaces: Oxymoron or binomial? Urban transformations in the digital era. *Geography Compass*, *13*(10), e12467.

de Waal, M., de Lange, M., & Bouw, M. (2020). The hackable city: Exploring collaborative citymaking in a network society. In: K. Willis & A. Aurigi (eds), *The Routledge companion to smart cities* (pp. 351–366). London and New York: Routledge.

Dean, J. (2008). Enjoying neoliberalism. *Cultural Politics*, 4(1), 47–72.

Di Vita, S., & Morandi, C. (2017). *Mega-events and legacies in post-metropolitan spaces: Expos and urban agendas*. Berlin: Springer.

Fan, T., & Lin, Z. (2022). Toward tourists–media–cities tourism: Xi'an as a Wanghong City. *Sustainability*, *14*(19), 11806.

Fed Square. (2017). *An Apple global flagship store to open at Federation Square*. http://fedsquare.com/an-apple-global-flagship-store-to-open-at-federation-square (page no longer exists).

Fed Square. (2018, May 25). *Hack Fed Square*. Fed Square. https://fedsquare.com/events/hack-fed-square

Fincher, R., Pardy, M., & Shaw, K. (2016). Place-making or place-masking? The everyday political economy of "making place". *Planning Theory & Practice*, *17*(4), 516–536.

Friesinger, G., Grenzfurthner, J., & Ballhausen, T. (2010). Urban hacking. *Culture Jamming Strategies in the Risky Spaces of Modernity*, Bielefeld: Transcript.

Garrett, B. L. (2013). *Explore everything: Place-hacking the city*. London: Verso Trade.

Graham, S., & Marvin, S. (1995). More than ducts and wires: Post-Fordism, cities and utility networks. In: P. Healy et al. (eds), *Managing cities: The new urban context*, 169–189.

Han, X. (2021). Historicising Wanghong economy: Connecting platforms through Wanghong and Wanghong incubators. *Celebrity Studies*, *12*(2), 317–325.

Harvey, D. (1985). *The urbanisation of capital: Studies in the history and theory of capitalist urbanisation*. Baltimore: Johns Hopkins University Press.

Harvey, D. (1990). Flexible accumulation through urbanization reflections on "postmodernism" in the American city. *Perspecta*, 28, 251–272.

Hesmondhalgh, D., & Baker, S. (2013). *Creative labour: Media work in three cultural industries*. London and New York: Routledge.

Huybrechts, L., Benesch, H., & Geib, J. (2017). Co-Design and the public realm. *CoDesign*, *13*(3), 145–147.

Johnson, P., & Robinson, P. (2014). Civic hackathons: Innovation, procurement, or civic engagement? *Review of Policy Research*, *31*(4), 349–357.

Kaethler, M., De Blust, S., & Devos, T. (2017). Ambiguity as agency: Critical opportunists in the neoliberal city. *CoDesign*, *13*(3), 175–186.
Kantor, P. (1987). The dependent city: The changing political economy of urban economic development in the United States. *Urban Affairs Quarterly*, 22(4), 493–520.
Logan, J. R., & Molotch, H. (2007). *Urban fortunes: The political economy of place, with a new preface*. Oakland: University of California Press.
Maalsen, S. (2021). The hack: What it is and why it matters to urban studies. *Urban Studies*, *59*(2), 453–465.
Maalsen, S., & Perng, S. (2016). Encountering the city at hackathons. *Code and the City, edited by R. Kitchin and SY Perng*, 190–199.
Marketing Entourage. (2019, August 30). *Hack Fed Square 2019-ME*. YouTube. https://www.youtube.com/watch?v=M5qxdFQtb2Y
Massey, D. (2005) *For space*. London: Sage.
Mattern, S. (2016). Interfacing urban intelligence. *Code and the City*, 4960.
Mattern, S. (2018). Maintenance and care. *Places Journal*. https://placesjournal.org/article/maintenance-and-care/
McQuire, S. (2017). *Geomedia: Networked cities and the future of public space*. Hoboken, NJ: John Wiley & Sons.
Nissenbaum, H. (2004). Hackers and the contested ontology of cyberspace. *New Media & Society*, 6(2), 195–217.
Osborne, S. P., Radnor, Z., & Strokosch, K. (2016). Co-production and the co-creation of value in public services: a suitable case for treatment? *Public Management Review*, *18*(5), 639–653.
Papanek, V., & Fuller, R. B. (1972). *Design for the real world* (p. 22). London: Thames and Hudson.
Perng, S. Y. (2019). Anticipating digital futures: Ruins, entanglements and the possibilities of shared technology making. *Mobilities*, *14*(4), 418–434.
Ping, L. (2024). Gentrification of Taste and Lifestyle: Shanghai as an Instagrammable Wanghong City. *The Developing Economies*, 2(1), 1–22.
Pogačar, K., & Žižek, A. (2016). Urban hackathon–alternative information based and participatory approach to urban development. *Procedia Engineering*, *161*, 1971–1976.
Qinghua University. (2018). *White Paper: Short Videos and City Image*. Centre of Qinghua University National Image Communications. Beijing: Qinghua University Press.
Redaelli, E. (2018). Creative placemaking and theories of art: Analysing a place-based NEA policy in Portland, OR. *Cities*, 72, 403–410.
Shaw, K., & Montana, G. (2016). Place-making in megaprojects in Melbourne. *Urban Policy and Research*, *34*(2), 166–189.
Sweeney, J., Mee, K., McGuirk, P., & Ruming, K. (2018). Assembling placemaking: Making and remaking place in a regenerating city. *Cultural Geographies*, *25*(4), 571–587.
Townsend, A. M. (2013). *Smart cities: Big data, civic hackers, and the quest for a new utopia*. New York: W. W. Norton & Company.
Tribe, J. (2018, February 14). *Apple and a perfect square. Herald Sun*. http://link.galegroup.com/apps/doc/A527370438/STND?u=unimelb&sid=STND&xid=f4d11931
Wang, V. (2023, May 21). Inside the Barbecue City that is China's hottest tourist destination. *New York Times*. https://www.nytimes.com/2023/05/21/world/asia/zibo-barbecue.html
Wark, M. (2006). Hackers. *Theory, Culture & Society*, *23*(2–3), 320–322.

Zapico, J. L., Pargman, D., Ebner, H., & Eriksson, E. (2013). Hacking sustainability: Broadening participation through green hackathons. In *Fourth International Symposium on End-User Development. June 10-13, 2013, IT University of Copenhagen, Denmark.*
Zhang, A. Y., Roast, A., & Morris, C. (2022). Wanghong urbanism: Towards a new urban-digital spectacle. *Mediapolis: A Journal of Cities and Culture*, 7(4), 1–18.
Zhu, W., & Chen, J. (2022). The spatial analysis of digital economy and urban development: A case study in Hangzhou, China. *Cities*, *123*, 103563.
Zuboff, S. (2019). *The age of surveillance capitalism: The fight for a human future at the new frontier of power*. London: Profile Books.

5 Governmental Assemblage

5.1 Introduction: Engagement in Digital Cities

Community engagement in Victoria, often conflated with civic or public engagement, is a deliberately planned process designed to foster active participation in decision-making. It differs from mere participation which often implies a sense of voluntary, as it involves a reciprocal exchange between individuals and institutions. Globally, public agencies have increasingly incorporated digital means to fortify the participatory governance and are faced with various opportunities and challenges (Spada & Allegretti, 2020). Especially in liberal democracies with existing engagement channels in traditional means, such as town halls, public forums, public hearings and consultation, digital engagement channels are often implemented to complement existing methods and increase citizens' opportunities to coproduce public decision-making (Osborne, Radnor & Strokosch, 2016; Smith, 2019).

In Australia, the engagement sector emerged gradually over the decades. Initially, citizen participation was primarily indirect, such as voting to be politically represented and political campaigning. However, the rise of community activism in the 1960s and local government reforms in the 1990s spurred the development of a more formalised engagement sector. The ascendance of New Public Governance (NPG) in the 2000s accelerated this trend, as governments sought to engage citizens in new and innovative ways. The institutionalisation of engagement practices has led to the emergence of a professionalised sector, characterised by dedicated staff, policies and frameworks. The Victorian Government, for instance, has implemented a range of policies, legislation and guidance notes to promote public engagement, including the Public Participation in Government Decision-making: Better Practice Guide, Victorian Auditor-General's Office (2015), Public Engagement and Partnership Vision (2018) and the Local Government Act (2020). Today, the engagement industry has developed into what engagement professional-turned-academic Helen Christensen (2021) described as a 'semblance of a professional sector'.

Three elements have become increasingly prominent: engagement policies, engagement professionals and engagement portals. Public engagement is now

DOI: 10.4324/9781003407973-5

legislated in some states like Victoria, requiring councils to conduct deliberative processes. Engagement professionals have emerged, driven by increased demand. These 'facilitators of democracy' (Christensen, 2021) can be government employees, and increasingly, public engagement is outsourced to external for-profit engagement consultants, who are found to improve cost-effectiveness while maintaining specialist skillsets (Christensen & Grant, 2020), despite potential negative impacts of market corruption (Hendriks & Carson, 2008). Industrial associations like International Association for Public Participation (IAP2) Australasia offer professional development and certification, potentially leading to regionally standardised practices. However, concerns exist about standardisation as it can displace local communities and human connections (Lee, 2014; McShane & Middha, 2021). Furthermore, government engagement portals have become the dominant platform for online engagement (e.g., Engage Victoria). These platforms, often using commercial engagement management systems, offer hybrid approaches that combine online and offline methods and are designed to address common challenges like time and budget constraints (Christensen & McQuestin, 2019). However, owing to processes of digitisation, globalisation and large-scale migration, the natural political constituents are increasingly disjointed with geographical demography. Dedicated engagement portals increase access while exacerbating the uneven distribution of socio-technical access.

Beyond Australia, in comparable liberal democratic settings, such as Norway and Spain, urban engagement and citizen e-participation also exist. Comparing and contrasting Oslo, Melbourne and Madrid, Legard and Hovik (2022) have argued that Madrid embraces the digital direct democracy model, as opposed to the crowdsourcing model in Melbourne and the e-bricolage model in Oslo, where citizens' involvement in the urban decision-making process is increasingly limited. East Asian governments are also increasingly recognising the strategic and competitive advantages of building robust national or regional digital infrastructure, developing technological tools and establishing digital ministries. The South Korean government, through the Minister of Science and Information and Communication Technologies (ICT), initiated a 'Digital New Deal' in 2021 to revitalise its economy post-pandemic. Japan's Digital Agency established at the cabinet level in 2021 aims to bridge digital divides and modernise government services by addressing the digital divide exposed by the Covid-19 pandemic. The Ministry of Digital Affairs of Taiwan was inaugurated on 27 August 2022 to integrate and develop policy for democratic engagement, information, cybersecurity and communication industries. These urban technical systems are what Winner (1980, p. 123) considered the 'inherently political technologies' that require or induce particular political relationships.

In my analysis, I draw on Federation Square Review material and Engage Victoria portal and engagement strategies from engagement professionals Capire and IAP2.[1] The Federation Square Review was commissioned by the Victorian Government and announced in April 2019, after Heritage Victoria rejected Fed Square Proprietary Limited's (FSPL's) application to demolish the Yarra Building. I investigate the background, process, results and implications of the Federation Square Review. I argue that the Federation Square Review successfully engaged a wide spectrum of relevant parties among lay and expert participants and led to structural changes committed to by the Victorian Government. However, as the Apple Federation Square proposal has shown, contemporary placemaking issues are no longer bound by state borders as sovereign issues or internal affairs. The review failed to reveal and respond to this trend that could be described as platform urbanism. I also discuss the Taiwanese g0v initiative and its implications, calling for a broader comprehension of digital placemaking, especially the importance of acknowledging the role of the governmental assemblage in digital placemaking initiatives as custodians of the processes with strategic foresight rather than reactionary agenda setters and solutionists.

5.2 Apple Town Square vs. Federation Square

The Federation Square Review emerged as Federation Square faced 'a period of change and challenge' (Federation Square Review Background Information, 2019). Though the Victorian Government's statements avoided mentioning 'Apple', the Apple controversy acted as a significant catalyst for the Federation Square Review. Apple's town square plan and the origin of Apple Federation Square provide crucial context for understanding the review.

5.2.1 Apple Town Square (ca. 2016–2019)

Spearheaded by then-retail chief Angela Ahrendts, Apple's short-lived town square plan aimed to rebrand its retail experience. Inspired by Starbucks's success in creating a 'third place' (Oldenburg & Brissett, 1982), Apple envisioned a hybrid retail store and educational centre offering workshops in various tech and creative fields. The goal was to transform Apple stores into community hubs, as Ahrendts expressed: 'Starbucks figured it out... "Meet me at Starbucks"... I'll know we've done a really, really great job if the next generation says, "Meet me at Apple. Did you see what's going on at Apple today?"' (Leswing, 2017).

This vision coincided with 'Today at Apple', a program of workshops and cultural events launched in US stores and later rolled out globally. Ahrendts even referred to stores as the company's 'largest product', with the architecture of the stores as the hardware and programming, 'Today at Apple', as the software.

Apple proposed several global town squares, including Stockholm, Washington, D.C. and Melbourne. Only the Washington, D.C. location, housed in the 114-year-old Carnegie Library building, opened successfully to the public in 2019. Public opposition led to rejections in Stockholm and Melbourne. The Stockholm City Council vetoed Stockholm's proposed location traditionally used for public events from Pride parades to election debates after public outrage. Similarly, the Melbourne proposal faced resistance (discussed previously). When Ahrendts left Apple in 2019, the town square concept lost momentum and vanished from official communications. However, 'Today at Apple' events continued in many Apple stores globally, including Melbourne.

5.2.2 A Controversial of Placemaking Bid

The Age, Melbourne's leading 'broadsheet' newspaper known for investigative journalism, revealed that the Apple plan for Federation Square underwent government analysis and was internally approved in mid-2016 (Lucas, 2018).[2] A confidential government report obtained by *The Age* showed a positive design review for the project completed in May 2016, along with plans to involve the square's original architect, Donald Bates. This timeline aligns with the rumours reported by the activist group, Citizens for Melbourne (see Chapter 3), previously reported by *The Age* in 2016. The Victorian Government only confirmed the proposal in December 2017.

Further investigation by *The Age* uncovered internal disagreements within the Labor cabinet. Three senior ministers—Planning Minister Richard Wynne, Creative Industries Minister Martin Foley and Special Minister of State Gavin Jennings—argued against the project but were overruled. Minister Wynne was then obligated to approve the plan and alter planning regulations to facilitate development, bypassing the City of Melbourne (COM) Council (Millar & Lucas, 2018). This arrangement has widely angered councillors, planners, architects and the public.

Lord Mayor Robert Doyle was briefed only shortly before the announcement and left office shortly after due to unrelated allegations. No other COM officials, including the Planning Chairman and former Labor Party State Secretary Nicholas Reece and prominent urban designer Rob Adams, were consulted about the project or the planning changes (Millar & Lucas, 2018). Following the announcement, the COM Council formally opposed the Apple plan. Councillor Rohan Leppert (of the Melbourne Greens) promptly moved a revocation motion that called on the Victorian Parliament to overturn Wynne's planning rule changes during the February 2018 Council meeting. The motion was unanimously supported by the COM but failed in the Parliamentary process (Leppert, Interview, 11 June 2019).

The following observations concerning the Apple town squares derive from the above accounts. First, albeit controversial, Apple positioned its town squares as an extension of placemaking initiatives. Ahrendts' comments about the store's hardware and software resonated with the concept of creative placemaking, highlighting the new retail strategy of using brick-and-mortar stores as media channels to amplify brand-building and deliver experience (Stephens, 2017). The site of Federation Square was not a random selection but a deliberate choice as Apple refused the Victorian Government's offers to establish a 'town square' in other Melbourne locations. This demonstrated that Apple valued highly the 'placemaking dividend' (McMahon, 2010, p. 16) provided by a public space like Federation Square. The place history, community, surroundings, walkability and attraction to local residents and international visitors alike would help Apple to attract more customers who would stay longer, spend more and revisit more frequently.

Second, Apple town square was exempted from and bypassed a commonly agreed urban planning process. As a global tech giant, Apple directly approached the Victorian Government, sidelined the COM and ignored the public consultation processes. Councillor Leppert described a top-down and ad hoc approach, with Apple seeking minimal government approval before navigating anticipated public resistance (Leppert, Interview, 11 June 2019). This strategy, coupled with the government's use of Cabinet confidentiality and commercial sensitivity to shield the deal, fuelled public anger. Apple's actions and the government's compliance contradicted the widely accepted principles of the public's right to know and participate in urban public infrastructure projects.

The controversy also reflected the complex interplay of urban and party politics in Victoria. Internal disagreements within the Labor cabinet, particularly with the Minister for Planning, who reportedly showed a lack of support for the Apple Federation Square, likely increased the probability of Heritage Council's intervention. The opposition Liberals criticised the project's failure to public consultation (Lucas, 2018) but voted with Labor to veto the Greens disallowance motion, according to an unspoken agreement to exchange future planning approvals. Ministers Foley and Wynne's opposition to the Apple plan was interpreted as an attempt to consolidate their inner-city seats from the Greens in the state election year of 2019 and prevent 'the potential political fallout from the perceived commercialisation of a public space' (Millar & Lucas, 2018, n.p.). The Apple plan further strained the relationship between the Victorian Government and the COM, previously partners in developing Federation Square in the mid-1990s.

The situation was rife with contingencies. It is unclear if the COM's official stance would have remained the same with a different mayor. The new lord mayor, Sally Capp, officially supported the Apple Federation Square plan. Although she could not have an official opinion on the matter due to a documented conflict of interest, she might have influenced the dynamics. However, all these politics and manoeuvres preceded public notification and discussion.[3]

5.3 Portal, Process and Protocol

The Federation Square Review, initiated in April 2019, aimed to ensure that the key community asset of Federation Square remained attractive and appealing. Overseen by a dedicated project team from the Victorian Department of Jobs, Precincts and Regions (DJPR),[4] the review unfolded in two streams.

The first stream, led by Jill Garner (Victorian Government Architect) and assisted by Rob Adams (Director of City Design at the COM), focused on public purpose and built form. It involved extensive community consultation conducted through the Victorian Government's online engagement portal and on-site surveys and in face-to-face group consultations led by Melbourne-based engagement consultants, Capire Consulting Group (Capire). This stream aimed to understand past and future public use of the square, develop a shared vision for its evolution and establish design principles and heritage control while considering upcoming developments. The second stream, led by Tim Eddy (former Managing Partner of Ernst and Young), addressed Federation Square's financial sustainability and governance. It examined the square's current and future financial capabilities, explored sustainable operating models and evaluated existing governance structures. Best practices in global public square management were also considered.

The results of the two streams informed a report with recommendations presented to the government in early 2020. The review process reportedly involved collaboration between the review chairs and the Federation Square's Board of Directors (BOD). While details about the second stream's progress were limited, the next session presents an analysis of the engagement process and portal affordance.

5.3.1 *Engage Victoria: A Platform for Public Input*

The Federation Square Review utilised Engage Victoria, an online platform launched in 2016 by the State of Victoria through the Department of the Premier and Cabinet for public consultation. Engage Victoria offers the public a central pathway for citizens to be involved in government decision-making. The Victorian departments and agencies can propose to host a consultation on Engage Victoria by completing an online form. Engage Victoria will forward the requests to a respective departmental engagement officer for further project requirements (Victorian Government, n.d.). Engage Victoria operates as a free website supported by Harvest Digital Planning, a Melbourne-based civic participation company. The public can participate by going to the website, completing surveys or leaving comments. They can also follow the project updates by creating an account and signing in on the website. Harvest Digital Planning also moderates the content on Engage Victoria on behalf of participating government agencies.

The platform displays a simple structure with a menu linking to the 'Home', 'About Us', 'Consultation' and 'Contact Us' pages. The 'Consultation' page lists ongoing and past projects. As of October 2021, there were approximately 40 open consultations, 300 updated consultations and 340 past consultations. New users can register, log in and follow projects for updates, but participating in open consultations does not require registration. Searching within the website is also enabled for current and previous consultations. Each consultation project has a dedicated webpage with similar layouts.

For the Federation Square Review, the webpage included an 'About the Review' section with background, a timeline, additional information and relevant links to external resources. 'Key Findings' were updated regularly to communicate the review procedures and results. A 'Document Library' archived key documents like the terms of reference, background information, stage reports and final reports. The following section provided varying tools that enable community sharing and exchanging ideas. For the Federation Square Review, a detailed survey, digital Post-it notes and a digital map were provided. The anonymous survey (10 minutes) asked quantitative and qualitative questions on the importance of good public spaces, the reasons for visiting, visitor experiences and future changes. The publics were prompted to rate whether they deemed Federation Square to be unique, safe, easy to navigate and a good place to visit. The digital Post-it notes modelled a creative brainstorming process similar to the fluorescent stickers used during Hack Fed Square. An interactive map allowed the public to drop pins on a specific location and share their opinions and recommendations.

The Federation Square Review employed a multi-faceted public engagement strategy, mobilising a diverse range of participants. The Community Consultation Report (2019) details that stage 1 has garnered input from approximately 1,700 individuals, 278 on-site surveys, over 400 on-site discussions with the locals and visitors, 621 online surveys, 389 online Post-it ideas, 63 map pins and 40 emails (Federation Square Review Community Consultation Report, 2019, p. 6). People between the ages of 20 and 74 participated from metropolitan Melbourne (83%), outer Melbourne (12%), interstate (3%) and overseas (2%), with the majority being female (57%; 41% male and 2% other). Initial findings indicated a strong public desire for Federation Square to 'reassert its status as Melbourne's civic and cultural centre' (p. 11). Many people suggested better connections to the Yarra River, more greenery, more sitting areas, better weather protection, smaller architectural changes and more cultural activities (Federation Square Review Community Consultation Report, 2019, p. 11).

Capire further conducted 36 interviews with key stakeholders, including 'the CEOs of Federation Square's cultural and arts tenants; food and beverage owners; leaders from the festivals, tourism and events sectors; heritage stakeholders; community interest groups; and representatives of key government agencies' (Federation Square Review Community Consultation Report, 2019, p. 12). Initial stakeholder feedback included determining Federation Square's strategic role and vision, calling for a more transparent governance structure and requesting a more sustainable business model and a more comprehensive understanding of the visitor experience.

The initial result of stage 1 significantly exceeded the participation levels in similar initiatives on Engagement Victoria. The report attributes this success to the combination of online and on-site engagement methods. Online participation allows access for geographically dispersed individuals and tends to attract younger and more informed groups (Delli Carpini, 2000; Weber et al., 2003). Conversely, on-site engagement provides access to those less familiar with online participation and offers flexibility in time commitment (Capire, 2016). The hybrid approach combats the common barriers of time, budget and technological limitations faced by public engagement initiatives (Christensen & McQuestin, 2019). The stage 1 report reflects a well-distributed participation overview, highlighting the success of this approach.

Dedicated portals as urban digital infrastructures create new urban political arrangements (Dunn, 2020) as they mediate the social processes of urban management and increase administrative efficiency, transparency and accountability (Coleman, 2009; Ransbeeck, 2020). Government engagement portals, like Engage Victoria, provide a targeted and centralised way for direct citizen–government interaction, partially avoiding the drawbacks commonly associated with commercial platforms and social media, such as privacy breaches and lacking direct dialogical processes (Piccorelli & Stivers, 2019). I did not observe tokenism associated with e-government's tendency to function as self-promotion and marketing (Medaglia & Zheng, 2017) without authentic citizen engagement (Piccorelli & Stivers, 2019). Engage Victoria was developed as a dedicated engagement portal without advertisement and other irrelevant information, reflecting the Victorian Government's attempts to increase participation, transparency and government accountability.

Further, unlike Hack Fed Square (see Chapter 4), stage 1 of the Federation Square Review focused on problem identification through open-ended questions rather than requesting specific solutions to a pre-formed problem. Indeed, the invitation for comments and questions will mean little if those inputs are not considered seriously or acted on as it was one part of a planning process. These questions, facilitated with digital Post-it notes and maps, tended to invite more site-specific and clarified answers rather than generalised responses. Although online engagement portals may not effectively address motivation issues outside of the technological domain, such as scepticism towards the government and fear of complex bureaucratic processes (Epstein et al., 2014), Engage Victoria fostered broader discussions and facilitated contributions from diverse public groups.

5.3.2 Stage 2: Public Consultation and Community Vision

In November 2019, one Capire consultant emailed me an invitation to participate in stage 2 of the Federation Square Review.[5] I participated in a consultation session with 14 others, including an Engage Victoria manager, Jill Garner, FSPL CEO Xavier Csar, FSPL representatives, Capire employees and several community members of various ages and backgrounds. The 90-minute consultation session was well structured, targeted and discursive, with a workbook guiding discussions on eight pre-determined questions. The manager from Engagement Victoria chaired the discussion of every question in the workbook, ensuring that all community members had the time to verbally express their ideas for all questions. Garner, Csar and other employees would ask questions for clarification and further discussion. We were encouraged to write our answers and thoughts on the workbook as the discussions went along. After all the questions, we handed in the workbooks with the option to get a digital copy of our answers. Two days after the consultative session, I received a movie voucher from Capire as a token of appreciation for my attendance.

The second stage of the Federation Square Review aimed to translate public input from stage 1 into a community-informed vision. It built on stage 1's results of local insights and tested the stage 1 findings to inform the development of FSPL's vision (Capire, n.d.). The design of the workbook reflected this aim. The first question asked for our response to the stage 1 findings. Questions two to five presented 14 statements about Federation Square's identity, the experiences on offer, commitment to FSPL's partners and FSPL's management style. We were asked first to agree or disagree with the statements and give comments. We were also asked to prioritise top visions for the square's future and identify potential indicators of success in achieving that vision.

Scholars like Cardullo and Kitchin (2019) position citizen-informed vision as a pinnacle form of citizen power in urban planning. However, my experience of collective visioning raised critical question about how this power translates into urban policy and administrative practice. First, concerns exist about accessing and utilising the vast amount of citizen data gathered. How can these responses be effectively analysed and transformed from raw data into actionable knowledge that informs policymaking? Second, how are these citizen responses mobilised and prioritised? Is there a risk of dominance by specific groups or perspectives within the community? How is the dominance of one key group to be avoided? How do we address silence, confusion and non-participation? How can we support citizens to better understand the issues and possible solutions? If offering limited choices inevitably shapes and constrains responses and solutions, how do we identify further actions when decision-making is necessary and imminent?

My experience at the consultation session highlighted some limitations of the process. Notably, the workbooks presented the findings of stage 1 as a starting point, and the session itself didn't delve deeply into FSPL's financial and governance issues. The limited choices presented for the future vision, particularly without considering financial constraints, hindered my ability to imagine realistic solutions. For instance, the option of 'affordable hosting of world-class festivals' seemed unrealistic without the knowledge of FSPL's financial capacity. This case underscores that while community vision co-creation can be a valuable tool for informing solutions on various public issues like energy, environment and budgeting (Cappellaro et al., 2019; Gabrys, 2014; March & Ribera-Fumaz, 2017), the process can be imperfect. Greater trust and transparency from all participants, including public authorities and consulting firms, are crucial for a meaningful co-creation process. Furthermore, strategies are needed to address issues of silence, confusion and non-participation among certain segments of the community. How can we support citizens to better understand the issues and potential solutions? The Federation Square Review serves as a microcosm of the challenges and opportunities inherent in citizen-informed visioning processes. While such processes hold promise for fostering a more democratic and inclusive approach to urban planning, careful consideration needs to be given to ensuring the accessibility, transparency and inclusivity of these exercises.

5.4 Evaluating Review Results

5.4.1 Federation Square Findings and Recommendations

The Federation Square Review, culminating in a report delivered to the Victorian Government in July 2020, involved a broad range of stakeholders. These included Traditional Owners, managing entities, cultural and retail tenants, relevant government departments, transport bodies, local government, tourism stakeholders, festival and event organisers, community groups and the general public.

The review identified four major findings. First, the communities and stakeholders aligned regarding Federation Square's future, emphasising a more strategic and civic-minded role for Federation Square to prioritise community and cultural activities. This agreement translated to a focus on strengthening First Peoples culture, art and heritage; enhancing connection to the Yarra River; and fostering collaboration among cultural tenants like Australian Centre of the Moving Image (ACMI), National Gallery of Victoria (NGV) and the Koorie Heritage Trust. Second, the review acknowledged the unique and complex architecture and design of Federation Square, highlighting the unfulfilled ambition of the original design to better integrate with surrounding contexts, particularly the Yarra River. The review also identified the complex governance environment involving multiple landowners, managers and planning authorities.

Third, the review identified the challenges of Federation Square's self-funded business model. The square's diverse utilisation modes, complex space and asset maintenance have created ongoing financial pressures. The review crystallised the need for improved governance to better align with Federation Square's civic and cultural objectives, emphasising the need for the BOD to upskill with relevant expertise in civic, cultural and design fields. Last, the review identified Federation Square's three mutually influencing challenges: declining public appeal, misaligned governance and unsustainable business model (Federation Square Review Key Findings, 2020, p. 13).

The review recommended three major reform areas. The first was to adopt a new community-informed vision and create an exceptional visitor experience. The new objectives expanded the original site objectives outlined in the Civic and Cultural Charter and incorporated new aims to collectively acknowledge and celebrate the Traditional Owners of Victoria and to protect Federation Square's architecture. The review recommended a new approach to programming and visitor experience to counter the increasing commercial atmosphere with curated commercial offerings. Actions included realigning and rebalancing its civic, cultural and commercial programming and visitor experience; establishing a programming and activation board subcommittee; developing a community and stakeholder engagement plan; and creating integrated programs between FSPL and its cultural tenants.

The second reform area was to ensure that the built environment supports the community vision by proposing new design principles. These principles focused on removing physical and programmatic boundaries, connecting with surrounding precincts and embracing the Yarra River. Additionally, the review emphasised celebrating First Peoples cultures, revitalising public spaces, prioritising pedestrians and completing unrealised design elements. Specific actions included exploring built-form improvements, better coordination with adjacent areas and establishing a design advisory panel.

The last reform area was to implement effective governance arrangements to meet its civic and cultural objectives. Key actions included amending FSPL's Constitution and Civic and Cultural Charter[6]; expanding the BOD's skills; and establishing subcommittees and advisory panels, including First Peoples representation and external expertise in events, arts, culture, design and architecture; and transferring portfolio responsibility for Federation Square to the Minister for Creative Industries (Federation Square Review Key Findings, 2020, p. 22).

The Victorian Government has accepted all review's major recommendations and key initial actions. These included adopting the community vision for Federation Square's cultural and civic impact, strengthening connections to the Yarra River and the adjacent arts and sports precincts, adhering to design principle for future changes and incorporating Federation Square in the government's Creative Industries portfolio alongside other cultural and civic icons such as National Gallery Victoria, Arts Centre Melbourne and State Library Victoria (Engage Victoria, 2020). The Victorian Government also invested A$20 million ($13.4 million) for immediate asset maintenance and renewal works. FSPL has committed to use the funds in line with the recommendations from the review, introduce new wayfinding and lighting facilities and enhance events programming.

5.4.2 *The Federation Square Review as Public Consultation*

The Federation Square Review chose the wording 'consultation' as its primary public engagement method. While 'consultation' and 'engagement' are sometimes used interchangeably (Crase et al., 2005), consultation is often viewed as a less potent form of participation (Stewart & Lithgow, 2015). For Arnstein (1969, 1975), consultation was categorised as a form of tokenism where the citizens have limited impact on the final decisions and lack assurance of changing the status quo. Cardullo and Kitchin (2019) further contend that, even with social media tools, consultations can be shaped by the dominant interests of urban administrations.

However, the Federation Square Review was not a mere formality, nor were its results predetermined. Archon Fung (2006) identified three key dimensions of the mechanism of participation: who participates, communication methods and the link to policy decisions. I reflect on each dimension below.

First, the review involved a broad spectrum of stakeholders, as discussed in the previous section. This ensured presentation from those most affected by Federation Square's future changes. Engaging with non-traditional stakeholders like lay audiences, activists and advocacy groups helped identify concerns outside of Federation Square's daily operation. The combination of expert and lay audiences effectively utilised diverse knowledge and local competence to tackle issues that the communities may face issues they are concerned about, but lack the necessary problem-solving experience (Hart, 2008). The participants of the review differed from the public at large in a Habermasian (1991) sense. Nonetheless, the review was successful in positioning public participation 'in synergy with representation and administration' to yield desirable outcomes rather than create confrontation (Fung, 2006, p. 68).

Second, the review aligned with the Victorian Government guidelines defining public engagement as 'a planned process with the specific purpose of working with individuals and groups to encourage active involvement in decisions that affect them or are of interest to them' (Engage Victoria About, n.d., n.p.). The platform aligns with the Victorian Government Public Engagement and Partnership Vision (Engage Victoria, 2018) developed in collaboration with Victorian engagement practitioners. The vision provides an overarching framework for public engagement by clearly setting community and system outcomes. The Victorian Government released the Victorian Public Engagement and Partnership Framework (2021–2025) in 2021. This framework further clarifies the principles, parameters and best practices of public engagement in the Victorian Public Service.

Third, informed by Arnstein's (1969) ladder model, the review observed the public engagement protocols developed by IAP2 of different levels of participation, including inform, consult, involve, collaborate and empower. Capire's principles mirrored IAP2, stressing clarity, acknowledging expectations, actively pursuing participants, openness, identifying organisations and leaders, being aware and self-reflective, respect and sufficient time (Capire, 2016). Different forms of public participation took place within the stage 2 consultation. Necessary information was provided. Deliberative discussions between participants have been encouraged. For instance, I had a small debate with one of the architects closely associated with the Our City, Our Square campaign who did not agree to allow more connection to the Yarra River. In my observation, nobody was purely a spectator in the session.

Last, the review culminated in significant structural changes that aligned with public and stakeholder interests. The first change involved respecting the decision of the Heritage Council of Victoria, acknowledging Federation

Square's historical, aesthetic, technological and social significance. This move overturned the previous Victorian Government's position and demonstrated a capacity for self-correction within a functioning liberal democracy. The second change involved transferring portfolio responsibility for Federation Square to the Minister for Creative Industries. This change meant that Federation Square would have access to recurring funding for its operations, similar to eight other state-owned creative organisations. These organisations combined hold cultural collections worth over A$5 billion ($3.4 billion) and support Victoria's creative industries through events, exhibitions, productions, networking forums, creative hubs, start-ups and research-and-development activities (Creative Victoria, n.d.).

Recognising Federation Square as a creative organisation reflects the governmental acceptance of the evolving nature of creative industries that are no longer limited to the production of traditional 'art and cultural' forms within enclosed institutional spaces (Papastergiadis, 2016). Creative industries typically encompass goods and services that produce cultural, artistic and entertainment value, such as publishing, visual and performing arts, TV, films, fashion and games (Caves, 2000; Flew, 2002). Australian states have historically prioritised funding for established art forms, such as opera, orchestral music, dance, theatre and circus (O'Connor, 2016). By placing an open public square alongside traditional art institutions (the GLAM [galleries, libraries, archives and museums] sector), the Victorian Government acknowledged Federation Square's contribution to the Victorian cultural scene and its value as a 'space of the curatorial' (Bauer & Oetker, 2018). This structural shift reflected a progressive understanding of urban culture and significantly altered FSPL's funding and business models.

While Arnstein's (1969) model suggested that public consultations often occur after key decisions have been made and actions already 'locked in', the Federation Square Review demonstrated a willingness to adapt based on community and stakeholder input. Furthermore, the review process did not reflect a binary confrontation between the mystified monolithic homogeneities of the 'have-nots' and the 'powerholders'. Instead, it involved navigating complex issues and collaborating among multiple actors in the unfolding of heterogeneous issues. Ultimately, the community, stakeholders and the Victorian Government discovered common ground through conflict, deliberation and collaboration.

However, my experience also highlighted ongoing challenges in public consultation practices. Examining the entire review process reveals aspects that remained semi-public. For example, stakeholder identification, the level of public participation and the consultation implementation details were not necessarily made public, for they might have been considered as 'internal' or 'confidential' to the government. Similarly, the separate stream that examines FSPL's business model and governance structure could be seen as a strategic decision made at the state level. These aspects require further consideration to ensure transparency and effectiveness in future public consultation.

5.5 Researcher Reflections on Urban Politics

Having accounted for the detailed processes and implications of the Federation Square Review, can we understand the review as an instance of the realisation of urban democracy? Existing scholarship associates urban democracy with three occasions: democratic elections and decentralised policymaking processes within a bounded locale, political struggles performed on the urban stage, and political struggles with the city as a political object (Lama-Rewal & Zérah, 2011). On the face of it, it seems the Federation Square Review has been quite successful in facilitating decentralised policy-making of Federation Square in the context of Victorian placemaking. However, seeing urban politics as an ongoing process, I contend that the Federation Square Review has three deficiencies. It lacks an open evaluation on its implementation, obscures the future arrangement for Federation Square as a state-owned creative organisation and fails to respond to the global scale of the Apple controversy. Similarly, some of the fast urbanisation and smart developments in Africa seek administrative autonomy and an 'extra-territorial status that enables property owners to assume the bureaucratic responsibilities and regulatory functions once reserved for exclusive control by municipal authorities' (Herbert & Murray, 2015, p. 475).

First, one satisfactory public review does not represent a universal model for the Victorian Government's engagement approach. Public engagement is inherently case-specific, often 'experimental' in nature, requiring adaptation to unique communities and their place based concerns. This sentiment aligns with the views of COM Digital Placemaking Lead Gemma Campbell, who emphasised the need for a nuanced approach. In an interview, Campbell used an interesting analogy:

> Public engagement is a bit like baking. You know, you're always balancing things. You measure them. Sometimes you need to sieve the flour. Sometimes you can't just bake the cake ... You have to just keep experimenting. This way, everything is in a balance ... There is no one formula, although we've got the [IAP2] spectrum. We understand when we go out to engage and we have our consultation recaps where we're explaining impact and where the community's feedback fits. But it also depends on the scale ... Full deliberative panels are incredibly time consuming and costly. A lot of the time, if you would ask the community to be involved in the process, they probably won't have the time.
>
> (Campbell, Interview, 2 June 2020)

Campbell highlighted the importance of respecting the specific context—the place and its communities—over a one-size-fits-all solution. The highest form of public empowerment and full deliberation can be time-consuming and impose unpaid labour on the communities, potentially hindering participation. Reflecting and evaluating the review with a broader professional community are crucial for ongoing improvement.

Despite Federation Square had secured ongoing funding, its status as a state-owned creative organisation ensures that its political complexities will persist. Access to recurring funding does not liberate the Square from the need to deliver its service in an economically measurable way. As Justin O'Connor (2016, p. 41) argues, viewing creativity primarily through an economic lens is unsurprising, considering that Victoria's cultural ecosystem operates 'in an age of corporate giants, secretive trade agreements, the symbiosis of real estate and planning and powerful telecom lobbies.' This entanglement suggests that the public goods nature of Federation Square and its cultural value will still be measured within the framework of urban capital production, circulation and distribution.

In June 2021, Creative Victoria announced the establishment of the Melbourne Arts Precinct Corporation (MAP Co) as part of the Melbourne Arts Precinct Transformation. This new entity would oversee both Federation Square and a planned 18,000-m^2 immersive public garden along St Kilda Road:

> MAP Co will take responsibility for the running of Federation Square, ensuring a coordinated approach to activating and maintaining these two landmark public spaces in the future. It will lead the delivery of the $1.4 billion Melbourne Arts Precinct Transformation project—Australia's largest ever cultural infrastructure project—working closely with project partners Creative Victoria, Development Victoria, National Gallery of Victoria and Arts Centre Melbourne. MAP Co's creation also responds to the 2020 Federation Square Review and will strengthen the square's governance and its cultural and civic purpose.
>
> (Creative Victoria, 2021)

While the announcement sparked no significant public debate or discontent, unlike the Apple proposal, further scrutiny is necessary, especially regarding the relationship between MAP Co and FSPL. Will the new entity adopt a constitution and charter similar to the FSPL? What will its economic and legal structure be? These unanswered questions highlight the ongoing political realities of managing urban public space.

Last, the Federation Square Review has failed to identify and respond to Apple's digital placemaking initiative, which was not a local event or an internal affair of the Victorian Government. Sarah Barns (2020, p. 15) argued in *Platform Urbanism* that the lines between physical and digital spaces are blurring, with software companies effectively becoming 'city builders'. An example is Google's subsidiary Alphabet's Sidewalk Labs, which aimed to integrate the digital intelligence of Google into reshaping the physical urban fabric. Additionally, platform companies such as Airbnb have emerged as new urban institutions that re-negotiate relations between market, state and community actors, for they can set the agenda of urban regulations and shape current and future policy debates pertaining to home sharing and city building (Van Doorn, 2020). Scott McQuire (2019) further highlighted how digital platforms have begun to organise certain urban logic by operationalising previously untapped public urban data for commercial gain.

Apple's vision for Federation Square differed from Alphabet's data-driven approach to smart cities. Apple entered the digital placemaking realm through the niche of the urban experience economy, leveraging Apple's expertise in product design and customer experience rather than data operations. Evidently, Apple hired Angela Ahrendts, former CEO of Burberry, to transfer an understanding of fashion to the product and shopping experience. Recalling how Ahrendts considered the Apple stores as Apple's largest product, Apple town squares aimed to package public spaces and urban experiences as products and integrate them into the Apple ecosystem by commercialising cultural and community building. Consumption becomes a seamless blend of tangible goods and experiential products. Sean Cubitt (2016, p. 82) aptly observes the shift in public aesthetics, where experiences are increasingly replacing goods due to the 'manic reproductivity of free-falling capital', requiring perpetual replacement in the cycles of capitalism.

The Apple controversy exemplifies the continued tension between global platformisation and urban governance. It can be viewed as part of a global civic movement against big tech companies seeking to transform urban public space. Other high-profile cases include public opposition to Alphabet's Sidewalk Lab in Toronto and Amazon's proposed new headquarters in New York City. Both cases ignited public concerns about surveillance cities and gentrification. The Federation Square Review's complete omission of the transnational dimension of digital placemaking highlights a need to understand these initiatives beyond local contexts and consider their broader political influence beyond the control of a sovereign government. The problematic Westphalian model of sovereignty, emphasising inviolable borders and non-interference, poses significant challenges for urban governments seeking to regulate transglobal tech companies and platforms.

The increasing inadequacy of liberal democratic structures to check the growing power of capital has led many scholars like Purcell (2002, p. 105) to argue that 'the hegemony of national-scale political and economic organisation is becoming unstable'. Lefebvre's notion of the 'right to the city' offered radical alternatives to these structures by rescaling the political community from national hegemonic to urban hegemonic (Purcell, 2002). However, establishing effective mechanisms for engaging urban inhabitants and global techno-capitalism in a meaningful conversation remains a critical challenge. Urban governments have a crucial role to play in facilitating this increasingly important dialogue.

5.6 G0v (Gov Zero)

Audrey Tang considered the below paragraph her 'job description' upon being appointed the Digital Minister in Taiwan in 2016 (Weyl & Tang, 2024):

When we see 'internet of things', let's make it an internet of beings.
When we see 'virtual reality', let's make it a shared reality.
When we see 'machine learning', let's make it collaborative learning.
When we see 'user experience', let's make it about human experience.
When we hear 'the singularity is near', let us remember: the Plurality is here.

Taiwan's digital strategy, inspired by three influential digital empires—the US, European Union (EU) and the People's Republic of China (PRC) (Liu, 2021)—blends decentralised innovation with human-centric governance and strategic technological investment. This hybrid approach positions Taiwan as a unique player in the global digital landscape, especially on two fronts: the appointment of a digital ministry and the proactive investment in civic hacker-inspired, digital engagement infrastructure. Following a prominent role in Taiwan's 2014 Sunflower Movement and the success in developing media literacy curricula for Taiwanese schools, Tang was appointed Minister without Portfolio responsible for open government, social innovation and youth participation to bridge the gap between older and younger generations in Taiwanese politics. In 2022, as a reorganisation of Taiwan's Executive Yuan,[7] a newly established Ministry of Digital Affairs was established to assume specific responsibilities from the previous Ministry of Transportation and Communications, the National Communications Commission, the Ministry of Economic Affairs and the Executive Yuan's Department of Cyber Security. Its remit encompassed the development of the digital economy and infrastructure, including government information security, digital services and data management, while fostering an innovative environment for technologically enhanced democratic advancement. Tang was appointed the inaugural Minister of Digital Affairs, succeeded by Huang Yen-nun in 2024.

Since 2012, Taiwan gradually established a series of digital civic infrastructure and institutions, most prominent of which are g0v, vTaiwan and the Taiwanese Presidential Hackathon. The ongoing community of G0v (https://g0v.tw/intl/en/) is simultaneously the symbol and the foundation of digital democracy in Taiwan. As an open-source, open-government collaboration, it was initiated by civic hackers like Kao Chia-liang as a response to opaque government data and deficient digital services through actions such as 'forking'—scraping data from '.gov.tw' websites and crafting user-centric interfaces hosted on their own platform (O'Flaherty, 2018). The unexpected popularity of these 'forks' spurred a fascinating convergence. Some government officials began to 'merge' these citizen-led innovations back into official services, blurring the lines between bottom-up citizen action and top-down state activity. vTaiwan (https://info.vtaiwan.tw/) emerged from a 2014 government hackathon upon the Minister without Portfolio Jaclyn Tsai's prompt for volunteers to create a platform for rational discussion and deliberation of policy issues that the entire nation could participate in. The vTaiwan process involves four flexible stages: proposal, opinion, reflection and legislation. This platform supports online and in-person consultation processes and has been used to crowdsource legislation related to the digital economy from alcohol sales to ride-sharing regulations. Despite struggles to sustain public participation and translate online engagement into tangible policy changes with lasting impact (Noveck, 2023), vTaiwan set a precedent of government and civil co-creation, leading to subsequent reforms like the Join platform in 2015 and participatory budgeting processes in 2016. The Taiwanese Presidential Hackathon offers a compelling model of public–private collaboration, galvanizing citizen innovation to address infrastructure challenges. This annual event, initiated in 2018, leverages quadratic voting to empower citizens in shaping the agenda and prioritizing projects. The Civil IoT (Internet of Things) project, a successful outcome, exemplifies how grassroots initiatives can scale to national-level impact. This model provides valuable lessons for nations seeking to harness the power of citizen-led innovation.

Compared with the digital engagement institutions in Australia, the Taiwan example is prominent in three ways. First, setting up a designated digital minister and, later, the Ministry of Digital Affairs allows for a whole-of-government approach to tackle complex and socio-technical issues that cannot be addressed by the existing government department alone. An example achieved by the Taiwanese Ministry of Digital Affairs is the Fintech Development and Innovation and Experiment Act (the Sandbox Act), which took effect on 31 January 2018, to enable private fintech businesses to test financial technologies and innovations in a controlled regulatory environment under the regulator's supervision. In contrast, Australia's 2019 Commonwealth government restructuring under the Australian Liberal Party, which dissolved the Department of Communications and the Arts and folded communications into the Department of Infrastructure, Transportation and Regional Development, reflects a more backward understanding of communication as a physical commodity rather than a digital phenomenon on a societal level. This backwards-looking approach risks hindering Australia's ability to compete in the digital age as it will significantly limit the discussions to the administrative and implementation side of the government rather than generate impact on political and strategic levels.

Second, the Taiwanese model of citizen engagement empowers citizens to propose and set agendas that alert ground-up problems promptly, fostering flexibility and responsiveness to emerging challenges. This approach hinges on a fundamental principle: trust in the public. By granting citizens agenda-setting power, governments can harness the collective intelligence of society to address complex issues. This digital-first approach finds semblance to the highly reputed Madrid model with the establishment of the Decide Madrid platform in 2015, as both instances bear a close connection between the government and the activist movements, the Taiwanese Sunflower Movement and the Spanish Indignados. In contrast, Australian citizens are limited in their ability to set policy agendas in an agile manner, nor can they easily access other people's discussions or in-depth submissions. Australia's current systems, lacking transparency and often requiring costly and bureaucratic processes like Freedom of Information submissions, can hinder citizen participation and limit the potential for innovative solutions. Taiwan's open-access approach to government data and public opinion offers a compelling model for fostering transparency and democratic engagement.

Lastly, when faced with complicated public policy questions, the Taiwanese model prioritises building rough consensus rather than solving the specific problem at hand. Specific tactics such as augmented deliberation are adopted at the opinion stage. vTaiwan uses the machine-learning software, Pol.is, to produce a visually mapped 'opinion landscape', highlighting key viewpoints, areas of agreement and dissenting voices, incentivising nuanced and consensus-driven statements. Particularly, the bridging comments between opposing opinions will be used for the reflection stage. This methodology helps mitigate social media's tendency to entrench social divisions and create echo chambers undermining a sense of shared reality. It steered away from demagoguery or politicians providing straightforward answers to appease the crowd, as it focuses on building the capacity for the citizens to participate in democracy continuously without taking a toll on social cohesion. In contrast, the discussion and reflection process in the Australian engagement sector can be considered non-existent. As previously established, democracy is not an unambiguous virtue—it is only as ever effective as the participants, institution, processes, practitioners and education surrounding it.

5.7 Conclusion

In this chapter, I have analysed the Federation Square Review instigated by the Victorian Government in the wake of the Apple controversy. I focused on the reviews' background, process, protocols and results. I identified that the Apple plan had been intertwined with Victorian placemaking politics and party politics before it emerged as a public controversy. Examining the Victorian Government's engagement portal and two-stage review process led me to conclude that the Federation Square Review was not a tick-the-box exercise but instead offered real opportunity for public engagement. The hybrid review, occurring both online and on-site at Federation Square, maximised community reach and recorded high participation. The first stage encouraged open-ended questions from the communities, while the second stage formulated a new community-informed vision for Federation Square based on the stage 1 community responses. The review has resulted in two structural changes in line with the public and stakeholder wills. First, the architectural site of Federation Square is protected because of its heritage value, with the newly created design advisory board overseeing its future physical changes. Second, Federation Square was officially recognised as a state-owned creative organisation and provided with recurring funding from the Victorian Government. However, I also identified the shortcomings of the review, such as failing to recognise placemaking as a continued process subject to the influence of global tech industries.

This chapter has illustrated the Federation Square Review as a process rife with tension and collaboration. By documenting the dissonance and disjunction over the Apple controversy within the governmental assemblage, I have shown that the Victorian Government was not a monolith in opposition to public opinion. Similarly, I have documented how the review successfully mobilised and enlisted collaboration among a wide spectrum of relevant parties, including Indigenous communities, the public, experts, business collaborators and relevant government sectors. In this way, I have shown that a public consultation process, albeit imperfect, can lead to more fruitful results than Arnstein's (1969) original conception. This opens further discussions about how to effectively design and lead public participatory practices for future urban decision-making. Together with the examples from Taiwan, this chapter considers the governmental assemblage as collaborations among diverse participants capable of enacting structural changes. The governmental assemblage, viewed not as a monolithic entity but as a network with blurred boundaries and internal tensions, challenges the binary understanding of urban politics as binary and zero-sum confrontations between the 'have-nots' and 'powerholders'.

Notes

1 Since the Apple controversy was a recent issue at the time of the research, accessing the members of the Victorian Government to discuss the still-sensitive issue was not possible. For example, a former Minister for Small Business, Innovation and Trade, who drove the Apple deal, declined my interview request, citing Cabinet in Confidence reasons. However, I utilised my interviews with COM professional staff and councilors who have witnessed the process and can offer insight into the Apple controversy and government engagement portals.

2 Efforts to interview a ministerial proponent of the Apple deal were unsuccessful. Owing to the 30-year confidentiality rules of Cabinet Records, details regarding proposal initiation and negotiation are unclear.

3 Chair of the FSPL's BOD Deb Beale had contributed to Capp's election campaign.

4 One of DJPR's functions has been to support Victorian tourism, sports and major event sectors, including Federation Square. See https://djpr.vic.gov.au/what-we-do/events-and-tourism.

5 Later, I found out from a Capire employee that I was contacted because my email was in FSPL's database after registering for Hack Fed Square (discussed in Chapter 4).

6 As of November 2024, no revised constitution or charter has been made available to the public.

7 The Executive Yuan is the executive branch of Taiwan. The head of the Executive Yuan is the Premier, the head of government, who has the power to appoint members to serve in the cabinet. The President, as the head of state, has the power to appoint the Premier and nominate the members of the cabinet.

References

Arnstein, S. (1975). Whom does the advocate planner serve (part two): Additional commentaries. In R. Cloward & F. F. Piven (Eds.), *The politics of turmoil: Poverty, race and the urban crisis* (pp. 54–55). New York: Vintage Press.

Arnstein, S. R. (1969). A ladder of citizen participation. *Journal of the American Institute of Planners*, *35*(4), 216–224.

Barns, S. (2020). *Platform urbanism: Negotiating platform ecosystems in connected cities*. Singapore: Springer.

Bauer, U. M. & Oetker, B. (eds.) (2018) *SouthEastAsia: Spaces of the Curatorial*. Berlin: Sternberg Press.

Capire. (2016). *Inclusive community engagement toolkit*. Melbourne: Capire. Version Two.

Capire. (n.d.) *Federation Square Review*. https://capire.com.au/projects/urbanandregionalplanning/federation-square-review/

Cappellaro, F., Chiarini, R., Meloni, C., & Snels, C. (2019). Smart community co-creation: The case of Centocello Project. *International Journal of Sustainable Energy Planning and Management*, *24*, 155–162.

Cardullo, P., & Kitchin, R. (2019). Being a 'citizen' in the smart city: Up and down the scaffold of smart citizen participation in Dublin, Ireland. *GeoJournal*, *84*(1), 1–13.

Caves, R. E. (2000). *Creative industries: Contracts between art and commerce* (No. 20). Cambridge, MA: Harvard University Press.

Christensen, H. E. (2021). Managing divided loyalties in the emerging profession of community engagement. *Journal of Sociology*, *57*(2), 196–212.

Christensen, H. E., & McQuestin, D. (2019). Community engagement in Australian local governments: A closer look and strategic implications. *Local Government Studies*, *45*(4), 453–480.

Christensen, H. E., & Grant, B. (2020). Outsourcing local democracy? Evidence for and implications of the commercialisation of community engagement in Australian local government. *Australian Journal of Political Science*, *55*(1), 20–37.

Coleman, S. (2009). Technological change and the shifting nature of political organization. In: A. Chadwick & P.N. Howard (eds). New York and London: *Routledge handbook of Internet politics*, 86–98.

Crase, L., Dollery, B., & Wallis, J. (2005). Community consultation in public policy: The case of the Murray-Darling Basin of Australia. *Australian Journal of Political Science*, *40*(2), 221–237.

Creative Victoria. (2021) *Cultural impact*. https://creative.vic.gov.au/research/creative-impact-framework/cultural-impact

Cubitt, S. (2016). Defining the Public in Piccadilly Circus. *Ambient Screens and Transnational Public Spaces*, 81–94.

Delli Carpini, M. X. (2000). Gen. com: Youth, civic engagement, and the new information environment. *Political Communication*, *17*(4), 341–349.

Dunn, P. T. (2020). Participatory Infrastructures: The Politics of Mobility Platforms. *Urban Planning*, *5*(4), 335.

Engage Victoria. (n.d.). *About*. https://engage.vic.gov.au/about

Epstein, D., Newhart, M., & Vernon, R. (2014). Not by technology alone: The "analog" aspects of online public engagement in policymaking. *Government Information Quarterly*, *31*(2), 337–344.

Federation Square Review Background Information. (2019). Federation Square Review Background Information. *Engage Victoria*. https://engage.vic.gov.au/download_file/18410/2787

Federation Square Review Community Consultation Report. (2019). Federation Square Review Community Consultation Report. *Engage Victoria*. https://engage.vic.gov.au/download_file/23311/2787

Federation Square Review Key Findings. (2020). Federation Square Review Key Findings. *Engage Victoria*. https://engage.vic.gov.au/download_file/31196/2787
Flew, T. (2002). Beyond ad hocery: defining creative industries. In *Cultural Sites, Cultural Theory, Cultural Policy, The Second International Conference on Cultural Policy Research, Wellington, New Zealand, 23-26, 2002.*
Fung, A. (2006). Varieties of participation in complex governance. *Public Administration Review*, *66*, 66–75.
Gabrys, J. (2014). Programming environments: Environmentality and citizen sensing in the smart city. *Environment and Planning D: Society and Space*, *32*(1), 30–48.
Habermas, J. (1991). *The structural transformation of the public sphere: An inquiry into a category of bourgeois society*. Cambridge, MA: MIT Press.
Hart, R. A. (2008). Stepping back from 'The ladder': Reflections on a model of participatory work with children. In: A. Reid et al (eds). *Participation and learning* (pp. 19–31). Dordrecht: Springer.
Herbert, C.W. & Murray, M.J. (2015) 'Building from scratch: New cities, privatised urbanism and the spatial restructuring of Johannesburg after Apartheid'. *International Journal of Urban and Regional Research*, *39*(3), 471–494.
Hendriks, C. M., & Carson, L. (2008) Can the market help the forum? Negotiating the commercialization of deliberative democracy. *Policy Sciences*, *41*(4), 293–313.
Lama-Rewal, S., & Zérah, M. H. (2011). Introduction. Urban democracy: A South Asian perspective. *South Asia Multidisciplinary Academic Journal*, (5), 1–17.
Lee, C. W. (2014) *Do-it-yourself democracy: The rise of the public engagement industry*. New York: Oxford University Press.
Legard, S., & Hovik, S. (2022). The impact of digital participation on democratic urban governance. *Citizen Participation in the Information Society*, 167–190.
Leswing, K. (2017, April 25). *Apple wants the Apple store to be the cool place for teens to hang out*. Business Insider. https://www.businessinsider.com.au/apple-wants-apple-stores-to-be-the-cool-place-for-teens-to-hang-out-2017-4
Liu, L. (2021). The rise of data politics: Digital China and the world. *Studies in Comparative International Development*, *56*(1), 45–67.
Lucas, C. (2018, February 12). *Apple plans for Fed Square were nearly complete in mid-2016*. The Age. https://www.theage.com.au/politics/victoria/apple-plans-for-fed-square-were-nearly-complete-in-mid-2016-20180212-p4z030.html
March, H., & Ribera-Fumaz, R. (2017). Against, for and beyond the smart city: Towards technological sovereignty in Barcelona. Association of American Geographers, Boston. April, 5–9.
McMahon, E. (2010). The place making dividend. *Planning Commissioners Journal*, *80*, 16.
McQuire, S. (2019). One map to rule them all? Google Maps as digital technical object. *Communication and the Public*, *4*(2), 150–165.
McShane, I., & Middha, B. (2021). Expectations and realities of digital public spaces: A case study of digital community engagement in Melbourne, Australia. *Articulo-Journal of Urban Research*, 22, 1–30.
Medaglia, R., & Zheng, L. (2017). Mapping government social media research and moving it forward: A framework and a research agenda. *Government Information Quarterly*, *34*(3), 496–510.
Millar, R., & Lucas, C. (2018, January 4). *Daniel Andrews' cabinet divided over Apple at Fed Square*. The Age. https://www.theage.com.au/national/victoria/daniel-andrews-cabinet-divided-over-apple-at-fed-square-20180104-h0dhaz.html
Noveck, B. S. (2023, November 23). *Was vTaiwan such a big flop, after all?* Medium. https://bethnoveck.medium.com/was-vtaiwan-such-a-big-flop-after-all-d6b365f916dc
O'Connor, J. (2016). After the creative industries: Why we need a cultural economy. *Platform Papers*, *47*, 1–60.

O'Flaherty, Kate (2018, May 4). *Taiwan's revolutionary hackers are forking the government.* Wired UK. https://www.wired.com/story/taiwan-sunflower-revolution-audrey-tang-g0v/
Oldenburg, R., & Brissett, D. (1982). The third place. *Qualitative Sociology* 5, 265–284.
Osborne, S. P., Radnor, Z., & Strokosch, K. (2016). Co-production and the co-creation of value in public services: A suitable case for treatment? *Public Management Review*, *18*(5), 639–653.
Papastergiadis, N. (Ed.). (2016). *Ambient screens and transnational public spaces.* Hong Kong University Press.
Piccorelli, J. T., & Stivers, C. (2019). Exiled to main street: How government's use of social media diminishes public space. *Journal of Public Affairs*, *19*(4), e1955.
Purcell, M. (2002). Excavating Lefebvre: The right to the city and its urban politics of the inhabitant. *GeoJournal*, *58*(2), 99–108.
Ransbeeck, W. V. (2020. May 22). *What's the difference between Civic Tech and GovTech?* https://digileaders.com/whats-the-difference-between-civic-tech-and-govtech/
Smith, G. (2019). Reflections on the theory and practice of democratic innovations, In S. Elstub & O. Escobar (Eds.) *Handbook of democratic innovation and governance.* Cheltenham, UK; Northampton, MA: Edward Elgar Publishing.
Spada, P., & Allegretti, G. (2020). When democratic innovations integrate multiple and diverse channels of social dialogue: Opportunities and challenges. In M. Adria (Ed.), *Using new media for citizen engagement and participation* (pp. 35–59). Hershey, PA: IGI Global.
Stephens, D. (2017). *Reengineering retail: The future of selling in a post-digital world.* Vancouver: Figure One Publishing.
Stewart, J., & Lithgow, S. (2015). Problems and prospects in community engagement in urban planning and decision-making: Three case studies from the Australian Capital Territory. *Policy Studies*, *36*(1), 18–34.
Van Doorn, N. (2020). A new institution on the block: On platform urbanism and Airbnb citizenship. *New Media & Society*, *22*(10), 1808–1826.
Victorian Government. (n.d.) *How to Set A Public Consultation.* Government of Victoria. https://www.vic.gov.au/set-public-consultation
Weber, L. M., Loumakis, A., & Bergman, J. (2003). Who participates and why? An analysis of citizens on the Internet and the mass public. *Social Science Computer Review*, *21*(1), 26–42.
Weyl, E. G., & Tang, A. (2024). *Plurality: The future of collaborative technology and democracy.* New York: RadicalXChange.
Winner, L. (1980). Do artifacts have politics? *Daedalus, Vol. 109, No. 1, Modern Technology: Problem or Opportunity?*, 121–136.

6 Conclusion

Reassembling Digital Placemaking

6.1 Ouroboroi and the Apple Federation Square Controversy

Determining the instigator of the dialogue and the substance of discussions between Apple Inc and the Victorian State Government is elusive, as the historical records remain classified as 'Cabinet in Confidence' for three decades. Few anticipated the widespread opposition to the vision itself. Fewer still could have predicted that the Apple Town Square proposal would act as a Pandora's Box, revealing the intricacies of contemporary urban politics, where multiple and interwoven concerns, both established and nascent, are laid bare. While the ouroboros traces the trajectory of one matter of concern, I employ the ouroboroi to underscore digital placemaking as a multifaceted and relational domain where various 'matters of concern' emerge, mature, resolve, stagnate, evolve or dissipate. By recounting the Apple Federation Square from the perspective of the ouroboroi, I disentangle contemporary urban complexities and expose the limitations of a linear hierarchical participatory model and a binary political imagination.

The first concern pertains to urban infrastructure (see Figure 6.1), specifically the proposal to demolish the Yarra Building to make way for the Apple Town Square at Federation Square. Contemporary literature has recognised infrastructure as an entangled ecology of material, administrative and intellectual systems, including policy, governing conventions and social norms (Easterling, 2014). Public resistance surged when the State Government, disregarding consultation, announced unequivocally that part of Federation Square would be demolished for Apple. This issue galvanised a broad coalition of activists, including Citizens for Melbourne, their allies and supporters, Heritage Trust Victoria, the City of Melbourne, the media and broader networks. The infrastructural concern has been allocated to the Victorian heritage policy catchment and subsequently entered the established trajectory of urban heritage protection processes involving nomination, confirmation, public submission, public hearing and decision release.

The infrastructural concern has been thoroughly examined, performed, debated and acted upon by the public and an expanded notion of civil society. The infrastructural concern traversed the full political circle, from political-1

DOI: 10.4324/9781003407973-6

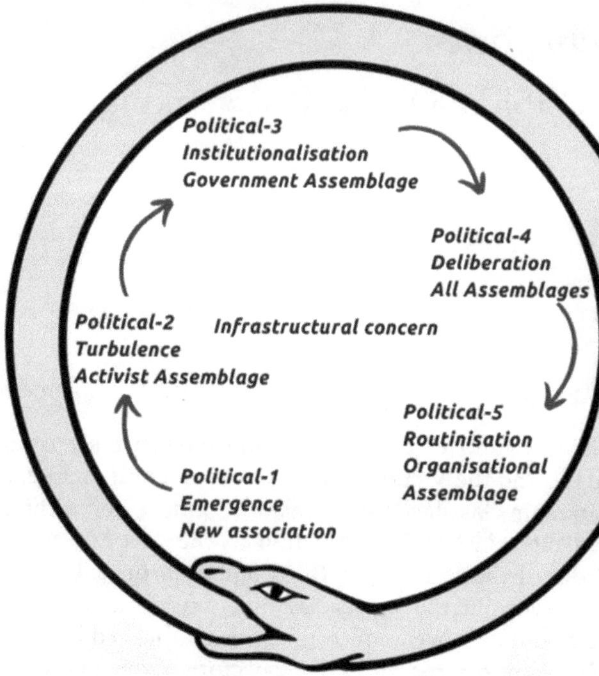

Figure 6.1 The Ouroboros of Digital Placemaking (Infrastructural Concern). Image by author.

to political-5. The Apple Federation Square proposal, as a new socio-technical association, threatened to alter the fabric of the Victorian public space (political-1), inciting public discontent (political-2). The Heritage Council solicited and reviewed public submissions (political-3 and political-4) and ultimately added Federation Square to the Victorian Heritage Register (political-5). The infrastructural concern was initially resolved with the recognition of Federation Square's heritage status, leading all involved parties, including Apple, to routinise and incorporate this new status into their daily operations.

The second concern relates to Federation Square's ability to reconcile its commercial objectives with its civic responsibilities in digital placemaking, specifically ensuring its sustainability while providing community value (see Figure 6.2). The absence of a sustainable business model precipitated the Apple controversy. Fed Square Proprietary Limited (FSPL) primarily self-financed the management of Federation Square as a public asset and the delivery of cultural programs. Apple was introduced as a commercial tenant to mitigate FSPL's financial strain. This model's viability was somewhat bolstered when the Victorian Government designated Federation Square as a state-owned cultural institution under Creative Victoria's governance and announced immediate capital funding. Consequently, this concern became enmeshed in the Victorian cultural policy cycle and governing trajectory.

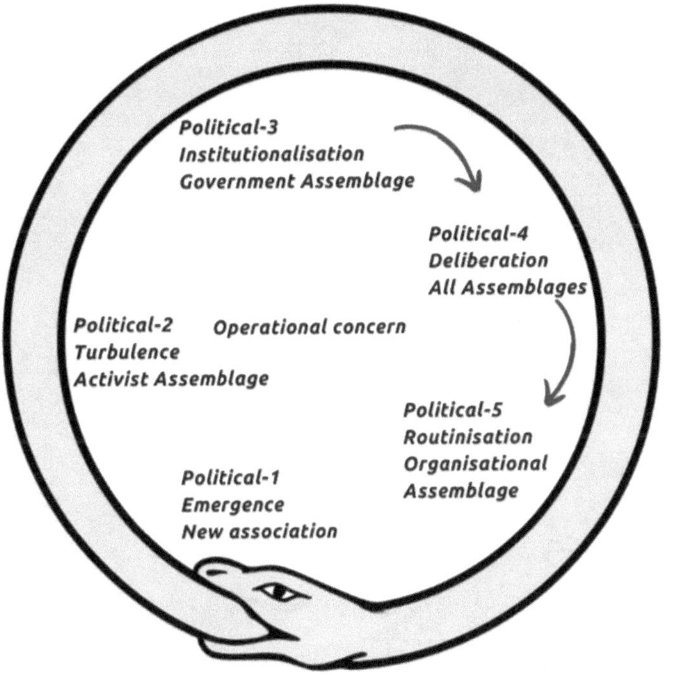

Figure 6.2 The Ouroboros of Digital Placemaking (Operation Concern). Image by author.

This operational concern, proven to be more intractable, has circumvented initial political scrutiny in political-1 and political-2. Most members of the public were unaware of Federation Square's financial deficits nor its need to self-fund its operations before 2020. This concern, perceived as 'mundane' and 'business as usual,' failed to capture the general public's interest and was confined to discussions among professionals, bureaucrats and academics. The issue surfaced formally during the Federation Square Review (political-3) and was addressed in stakeholder engagement sessions (political-4). Although Federation Square's identity crisis has been mitigated by its inclusion in the Creative Victoria portfolio, thereby securing ongoing funding (political-5), the operational challenge remains unresolved. The corporatisation and commercialisation trends within the shrinking Victorian arts and cultural sectors extended beyond Federation Square. With the establishment of the Melbourne Arts Precinct Corporation (MAP Co), this issue is likely to persist and evolve.

The third concern, still emerging, regards the capability of platforms and tech companies to transform urban socio-spatial structures (see Figure 6.3), namely the technopolitical impact of Apple on digital cities. This issue is denied by industrial actors, insufficiently acknowledged by governmental assemblages, and tentatively reflected upon by activist groups. However, this line of inquiry and the nascent questioning voice have dissipated over time.

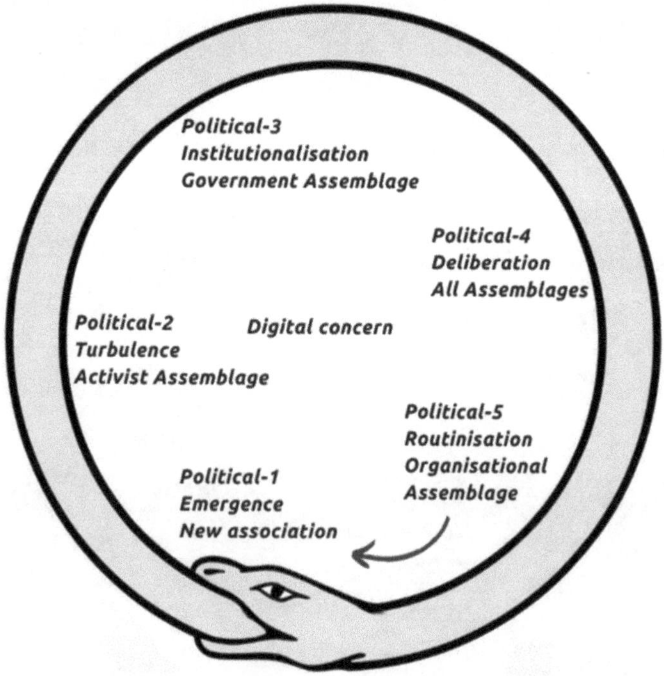

Figure 6.3 The Ouroboros of Digital Placemaking (Digital Concern). Image by author.

The extent to which tech companies shape placemaking industries, alter urban political processes and corporatise public culture remains underexplored.

The overarching concern about the holistic impact of digital technologies and platforms has not yet permeated the Victorian Government's agenda. Several federal reviews and inquiries have examined digital media production and platform impacts,[1] yet the urban environment and political processes have not been a focal point. It is noteworthy that digital solutions and smart city technologies are often taken for granted as ubiquitous urban realities and heralded methods of public administration (Zhao et al., 2021), without undergoing significant public scrutiny or debate outside the intelligentsia. The growing influence of what Kitchin (2023a) termed 'temporal doxa'—the valorisation of digital technologies as instantaneous solutions—precludes time for reflection or deliberation (Virilio, 1986). In other words, the market-driven innovation and speculation logic incentivise the industries to seek forgiveness rather than permission. In the context of the Apple incident, a few activists have partially recognised this emerging phenomenon and its potential to create new realities and transform urban power structures by generating digitally mediated time/space scapes (so I situate the digital concern between political-1 and political-2). This concern has yet to find a coherent vocabulary, policy framework or political trajectory. The Victorian

Government must acknowledge the new challenges posed by technologies and establish channels and processes for the public debate.

When discussing the brief tenure of Apple's town square initiative and Angela Ahrendts' departure from the company, one interviewee highlighted the contrast between Ahrendts' fashion background and the technological essence of Apple stores: Fashion events are transient, bringing together diverse elements for a limited time before dissipating. In contrast, Apple stores are physical, substantial and enduring (Davidge, Interview, 1 May 2019). This observation underscores how digital placemaking initiatives influence various urban temporal layers through both infrastructural and cultural manifestations. The impact of digital placemaking unfolds across multiple scales, spanning across the everyday, local, national, transnational and beyond. I have illustrated how global digital capitalism reshapes public spaces and urban social life through myriad 'microspatialities' (Dovey, 2011, p. 348) that coalesce into 'synergistic emergent effects at higher levels.' This conceptual reassembly elucidates the risks of reductively framing emergent phenomena like digital placemaking within political economic paradigms such as globalisation and neoliberal urbanism.

This Apple Federation Square case illuminates the divergent concerns within digital cities, each at varying stages of maturity and attracting distinct participatory cohorts. These three concerns have evolved uniquely within the context of Victorian politics. Urban planning in Victoria dates back to the British colonial period of the late 18th century, and heritage protection initiatives emerged in the 1950s. The state's first cultural policy was introduced in 2016, following Australia's inaugural national cultural policy, Creative Nation (Government of Australia, 1994). With historically robust institutional, business and civic supports, urban activist campaigns in Victoria have unsurprisingly advanced the furthest and achieved the most significant outcomes. Analogous to Latour's retelling of Louis Pasteur's triumph in winning over the support of the French farmers, industrialists, politicians and the scientific establishment, Citizens for Melbourne have outpaced Apple in 'amassing animate and inanimate allies' for their goals regarding the infrastructural concern (Harman, 2014, p. 45). In the Victorian context, architecture and cultural operations have been accepted as political subjects within established political institutions. When concerns are raised, they follow existing protocols for resolution (i.e., heritage and cultural policies). Conversely, the digital concerns remain nascent, necessitating the development of a lexicon for their articulation and governmental framework addressing digital concerns within the Australian polity. The absence of a dedicated government portfolio, responsible ministers and due process at both the commonwealth and state levels underscores a lack of strategic foresight in guiding a digital nation's trajectory. As of this writing, Australia has yet to establish an analogue to Taiwan's Ministry of Digital Affairs, instead linking technology portfolios closely with industry and growth and communications with transportation (see Chapter 5).

Given the ubiquitous and polymorphic nature of public participation, it defies measurement against a linear model of the urban power echelon. Participation is multifaceted and cannot be reduced to clear-cut categories. In a sense, urban activism, which disrupts the routine of digital cities, is as crucial as urban administration's mundane caretaking of public spaces. As Murphy (2015) argues, maintenance and care extend beyond the physical upkeep of urban objects; they involve negotiating social orders and shaping user behaviour within broader socio-technical and political frameworks. Similarly, the realisation of community visions hinges on their long-term viability and dynamism in the real-world surroundings. Public participation is not inherently antagonistic to commercial incentives or neoliberal frameworks but inevitably interacts with them. Thus, a collaborative reconciliation between community and commerce (Bruns, 2012) and a nuanced understanding of public power relative to the urban economy are key. The focus should shift from arbitrarily dismissing certain participatory activities as insignificant towards empowering citizens to recognise the world-changing potential inherent in their actions, regardless of their conventionally perceived significance. This reframing necessitates a rethinking of traditional power structures and a celebration of the diverse voices that contribute to the symphony of public participation in digital placemaking.

Examining the Apple Federation Square as an intertwining of urban socio-technical assemblages rather than entities is a process-oriented urban analysis, focusing not solely on the short-term outcomes of urban conflicts. Authority or power does not reside in any urban entity but is 'produced through associations made between socio and material agencies' (Schwennesen, 2019). As projectual dispositions, conceptual models and our narratives of them both obscure and reflect our perceptions, and they feed into the reality of power dynamics. Increasingly, urban scholars recognise that urban power transcends a simplistic binary of domination and oppression, imposition and opposition (Paddison et al., 2002); it is instead 'relational and entangled, often fragmented, uneven, inconsistent and paradoxical' (Kitchin, 2023b, p. 32). Nonetheless, discussions frequently fall into a dualistic framing of participatory justice versus political economic power (Kitchin, 2023b). As articulated in the introduction, the ouroboros of digital placemaking offers theoretical symmetry to participatory powers within the digital cities by fundamentally rejecting the dualistic framing.

In recounting the Apple Federation Square incident, I acknowledged the disparities between the 'haves' and 'have-nots' and the realities of conflicts, extraction and domination. However, I rejected a rigid narrative that frames the logic of hyper-capitalism as the 'context of context' (Brenner et al., 2011) for analysing contemporary cities. I did not abdicate moral responsibility to technological systems nor dismiss the potential for individuals to wield diverse forms of power and effectuate unplanned outcomes. I considered the governmental vision, entrepreneurial drive, civic volition and administrative routines integral to digital placemaking. Additionally, I accounted for the emergence of non-profit technical data systems, such as government engagement portals and civic hack initiatives, as digital placemaking evolves. Therefore, the Apple controversy cannot be simplistically reduced to a binary conflict between a failed business proposal and a public victory. Urban political processes extend beyond singular issues and interest groups, reflecting intricate and multifaceted dynamics. A process-oriented urban analysis is needed, so that the focus is not only on one-time results. Thus, the challenge lies in recognising the urban complexity without allowing it to serve as an alibi, balancing nuanced understanding with actionable insight.

6.2 Digital Placemaking: Continuity and Rupture

Contemporary academic discourse has shifted the focus towards examining urban transformations and spatial production amid digitisations. This discourse often oscillates between extremes: simplifying digital placemaking to digitally represent urban spaces and services at one extreme and an instrumentalist view that sees digital platforms as tools liberating us from physical complexities and risks at the other. However, conceptualising and governing digital places must transcend mere technical regulation. Contrary to Mark Zuckerberg's vision, a digital 'town square' should achieve more than a virtual container for opinion exchange. Adopting a relational and constructivist approach reveals profound changes in spatial production processes. As human geographer Tim Cresswell (2014) elucidated, a place is a meaningful site that amalgamates location (the geographic coordinates), locale (the material setting of social relations) and a sense of place (the emotional and psychological attachment). For instance, Pokémon Go exemplifies a digital place by fostering quests through collective urban exploration among gamers. A post-Taylorist Amazon warehouse is a digital place where workers share precarious conditions, biographical data, performance metrics and thwarted unionisation efforts. Similarly, design hackathons, virtual or in-person, represent digital places rooted in a digital ethos of decentralisation and agility, simultaneously being a community-building experience, a place-branding exercise and a compromise to urban entrepreneurial drives that many would interpret as neoliberal urbanism. While advancements like Apple's Vision Pro or digital twin modelling with enhanced visualisation capabilities, we are already inhabiting digital places regardless of the latest technological regimes. This nuanced understanding of digital placemaking underscores its multifaceted nature and the necessity to engage beyond simplistic regulatory or instrumentalist perspectives.

In this work, I have deliberately avoided suggesting that digital placemaking imposes radical transformations on urban configurations, as such a claim would constitute another mythologising and deterministic ideology. Similarly, this book refrains from dichotomising the specificity of place against the perceived abstraction of digitality, instead conceptualising digital placemaking as a holistic practice for inhabiting, thinking and acting. This perspective resists the tendency to over-philosophise places as mere nodes and vectors within global media, capital and power flows and instead views places as contextualised, open-ended practices and events, attentive to site-specific particularities. In line with Massey (2005), who argued that places are infinite and boundless and possess multiple identities, *Reassembling Digital Placemaking* shifts the focus from an empty, sedentary notion of place to the active processes of placemaking. Lewis Mumford (1937) conveyed that the city is a geographical plexus, an economic organisation, an institutional process, a theatre of social action and an aesthetic symbol of collective unity. The emphasis of this book is placed on the symbolic and socio-material realms, specifically the spatial practices within socio-technical contexts and ordering (Gillespie et al., 2014), with the unfolding and colliding of the activist, organisational and government assemblages.

Digital placemaking extends the legacy of traditional placemaking, evolving through the logic of assemblage as diverse stakeholders enter the digital urban landscape. It integrates with the contemporary technological milieu, situating itself within multifaceted urban trajectories and emerging as a dynamic site of contention and collaboration. This book delineates the development of digital placemaking, highlighting its incorporation into urban planning, design consultancies, government initiatives, community projects, artistic endeavours and the tech industries. This expansion reflects Marshall McLuhan's (1967) concept of the invisible environment, where digital technologies pervade our surroundings and operations, receding from consciousness and becoming taken for granted. In this way, the communitarian quality of digital placemaking conditions rather than stipulates moral imperatives. In this book, I do not judge whether digital placemaking is a 'good or bad' force but demonstrate the user-generated, participatory quality of digital placemaking is contingent on the prescribed socio-technical environments (Gabrys, 2014). I further demonstrated the socio-technical environments are consequently and subsequently undergoing transformations due to the emergence of new techne and praxis, underscoring Latour's (1990) famous assertion, 'technology is society made durable'. Martin Heidegger (1966) has cautions against technology's tendency to 'enframe' (Gestell) nature and humanity, reducing them to mere resources for extraction. He proposed releasement (Gelassenheit) as an antidote, achieved through transformative thinking as a remedy against the limitations of modern techno-scientific practices. Avoiding the technological deterministic pitfalls, this interconnectedness reinforces the tautological nature of digital placemaking, highlighting its intrinsic alignment with the prevailing technological frameworks that ceaselessly shape our understanding and interaction with digital environments.

Moreover, the action of reassembling situates digital placemaking at the confluence of variegated matters of concern, temporality and scale. The 'matters of concern are not firmly substantiated, prone to unintended consequences and deeply intertwined with particular situations' (Chhachhi & Donovan, 2014, p. 881). Reassembling digital placemaking prompts rethinking urban media beyond locality and immediacy against the abstract logic of modernism and the cynicism of postmodernism. The matters of concern entangled in digital placemaking initiatives are multiple and may exist beyond immediate public recognition or pre-existing government agenda. I have accounted for the infrastructural, operational and digital concerns of the Apple controversy, but digital placemaking can open to more concerns that 'destabilise and reorder our collectives' (Blok & Farías, 2016, p. 6) and invite more questions. For example, as multifaceted and dynamic processes as the symbiosis between technologies and urban spaces, digital placemaking creates a potential ethos for urban governance and organisation that challenges the dualities between democracy and technocracy. Digital placemaking can also potentiate a shift from an economy of growth to an economy of care and maintenance that continuously improves the delivery of goods and services based on real-time user feedback and market momentum. Since digital placemaking is increasingly interdependent on citizen sensing and data capturing, data flows in the digital cities are decentralised and multi-directional rather than merely from the city to the inhabitant (reminiscent of information flow in the broadcasting regime). Finding 'balance' in this relation remains one of the biggest challenges of contemporary urbanism and digital policy. Urban governments must recognise that legacy issues may arise alongside novel ones and enable public consultation to address these emerging concerns beyond existing policy frameworks.

Digital placemaking, as reconceptualised, emerges as a relational and dynamic locus of contention and collaboration among citizens, experts, industries and governments. This phenomenon, particularly prevalent in more affluent regions and cities, has proliferated and become routine within urban governance. Initiatives such as urban artivism, urban branding, design hackathons, public consultations and tech placemaking are neither purely top-down nor entirely grassroots but intersect in complex and unforeseen ways (Simone, 2010, p. 4). These participatory forms inject vitality and unpredictability into urban politics and must be understood as part of the assemblage that embeds them. Reassembling digital placemaking synthesises various strands of participatory discourse in urban and media studies, showcasing their multitude, power and contingency. Even seemingly trivial instances of public participation contribute to a larger continuum that profoundly influences political dynamics. The ouroboros metaphor underscores the importance of collaboration over domination and advocates for an intersectional perspective beyond strict industrial and institutional limitations to comprehend the diverse strategies, implementations and evaluations of participatory practices. Urban governance thus engages with a broader spectrum of public imagination and expression, oscillating between naïveté, nonchalance and populism. Local governments, often considered 'schools of democracy' (Mill, 1977; Sharpe, 1970), offer greater participatory opportunities than higher levels of government. Digital placemaking substantiates the meta-power game on the local level in which planetary capital, states and civil society are its players at an unprecedented speed.

6.3 Socio-technical Natality

Contemporary cities, burgeoning in their knowledge entrepreneurship, creative ambition and technological capacities, have increasingly invested in civic infrastructures such as open plazas, accessible public Wi-Fi, engagement portals and open-source government databases. These contested spaces provoke decentralised and situational engagement divergent from the classical modes of public participation in city-states or the Bourgeoisie public sphere cultivated in the mass media era. Whether publicly or privately funded or as public–private partnerships, these initiatives elude sole allegiance to technocracy or neoliberalism, adding new paradoxes to urban governance. On the one hand, new technologies engender novel urban forms, functionalities, political typologies and commercial establishments, concurrently fostering novel manifestations of community, participation and belonging. On the other hand, digital cities capture, amass and operationalise new data types—public attention, sentiment, behaviour and activity—and translate them into quantifiable metrics such as visitation numbers, heatmaps, geodata and place scores. Digital placemaking is in a perpetual state of defying its own product and device. It resists the rigid paradigms of master planning, computational abstraction and quantitative optimisation, honouring nuanced, embodied and interconnected urban realities with boundless potential to surprise.

For Hannah Arendt (1958), each new birth introduces a novel and unexpected beginning in the world. Every new human and nonhuman association possesses the potential to chart new trajectories. The spontaneity and unpredictability inherent in all beginnings hinge on a sense of 'hard-earned' freedom, realised only through systems of checks and balances. Freedom is the absence of neither labour nor politics. Contrary to the modernist reduction of freedom to the absence of political engagement, Arendt vehemently refuted the idea that liberty begins where politics end, a belief perpetuated by theorists like John Locke (1689), who equated freedom with the lack of political structures. While Herbert Marcuse (1964) contended that technological rationality could ideally eradicate the burdens of repetitive labour, technological progress may alleviate but never entirely dispense with such necessities. Acemoglu and Robinson (2020) further nuanced this discourse, positing that democratic freedom resides within a 'narrow corridor' between social collapse and authoritarianism, maintained through an ongoing struggle between state and society. Arendtian natality differs from Foucauldian biopolitics (Foucault, 2003) in that the latter depicts how the governmental power regulates and disciplines the human bodies and lives, whereas the former believes in a notion of power that is transient and emergent with the demos. Similarly, socio-technical natality arises in an ever-changing, complex system rather than a self-contained, totalising system where all parts support each other.

Digital placemaking emerges as a nuanced confluence of technological innovation and socio-political engagement, embodying a continual negotiation of agency within the urban milieu. In *Plurality: the Future of Collaborative Technology and Democracy*, authors Glen Weyl and Audrey Tang (2024) contend that the global society has only begun to envision large-scale digital democracy through public investments, research agenda and private development. This evolution compounds the challenges that democracies must address. Reassembling digital placemaking aspires to protect, care and take actions (Latour, 2004, p. 232) rather than mere debunking and critique. In recent years, diversity, equity and inclusivity (DEI) initiatives, sustainability and social justice have been incorporated into planning and placemaking discourse (Friedman, 2023; Saitta, 2024). Persistent inquiries into the transformative impact of digital technologies on participatory parameters remain pressing (Levenda et al., 2020). According to Arendt (1961), an ideological mindset results in purely logical deductions that begin with a predetermined premise and proceed through abstract reasoning to be applied to all aspects of reality, thereby overlooking the intricate complexities of the real world. In this sense, the ouroboros of digital placemaking is not conceived through ideological reasoning nor aims to produce ideological weaponry. Rather, it seeks to propose an agenda to better implement the cases of digital placemaking, focusing on effective and inclusive processes, innovative and alternative methods and analysis and translation into policy.

Effective and inclusive processes

- How can the communities be included in agenda setting and decision-making of public affairs?
- How can diverse public opinion be effectively gauged and enacted in a data-rich environment?
- How do we match the matters of concern and the legitimate scope of public to be mobilised?
- Who should spearhead engagement sessions—activists, place managers, governments, third-party experts or tech companies?
- How should these initiatives be funded, and should participants receive remuneration?

Innovative and alternative methods

- What alternative tools and participatory methods can attract typically disengaged groups or identify excluded communities?
- What data should be generated and collected to facilitate this process, and what privacy and legal considerations are paramount?
- Can public consultations that embrace play, artistic and creative expressions be equated with traditional consultation processes used by urban governments?

Translation and implementation into policy

- How does one ensure the representativeness and feasibility of the public engagement results?
- How does one ensure the results create social cohesion rather than deepening gaps in society?
- How should the outcomes be translated into policy, and how can these results be communicated to a broader audience?
- What protocols should govern these new forms of collaboration, and who will mediate the process?
- How will the findings be implemented in a sustainable manner rather than as a tick-a-box exercise?

Digital placemaking signifies a paradigm shift from the era of data scarcity to one characterised by data abundance. Shoshana Zuboff (2019) elucidates this transition by differentiating surveillance capitalism from traditional market capitalism. For Zuboff, surveillance capitalism, unlike its predecessor, dispels mystery with certainty, fundamentally altering the market's classical notion of inherent unknowability. The concept of the 'Big Other' encompasses pervasive technological systems equipped with indelible memories and the capacity to distil abstract patterns, transforming previously unknowable market dynamics into realms of predictability and control. However, it is imperative to recognise that data does not equate to knowledge, value, affect, ethos or truth. This realisation prompts critical inquiries about the technological paradigm of digital placemaking: How can we discern knowledge from noise and disinformation? How can we prevent the logic of abstraction from erasing nuanced realities? How can we channel technology to care rather than control and how to use technology for creating nuance rather than novelty? Moreover, how do we reimagine and reconstruct the 'space of appearance' as conceptualised by Arendt (1958), in an era where the images of contemporary cities are digitally mediated and technologically saturated (Rose, 2022)?

In delineating the emergent paradigm of digital placemaking, I propose three foundational principles central to understanding a new mode of sociotechnical natality intrinsic to digital places. Socio-technical natality signifies a transformative genesis, wherein new beginnings are collectively orchestrated by the synergistic interplay between human agency and technological advancements, in a way the scientific is indistinguishable from the political. This collaborative dynamic emancipates innovative potentialities, fostering environments where social and technological forces converge to co-create digital places imbued with novel meanings and possibilities. Within this paradigm, accountability in digital urbanity is reimagined as networked, distributed and non-human-centric. Responsibility surpasses the monopolistic limit of entities like the 'Big Other' or technocratic figures such as Jeff Bezos; rather, it is embedded in the myriad actants, actions and interactions that collectively engender meaningful change. As media technologies, long understood in their capacities to overcome spatiotemporal limits and implicated in placemaking activities, the city of tomorrow is already being conceived, visualised and built today, disrupting the linearity of time. The once mythologised notion of the future as a radical departure from the present is replaced by 'participatory futuring' (Kitchin, 2023a), which crafts an inclusive, preferred future by engaging in normative planning exercises that anticipate, apprehend and redirect potential future trajectories.

First, empower citizens to develop and advance new socio-technical competencies. In the book *Urban Commons*, Daniel O'Brien (2018) elucidates how Boston's 311 public reporting system underscores the imperative of cultivating these skills to optimise urban governance and civic accountability. Initiatives and discourses such as the new urban science (Townsend, 2013), citizen science (Vohland et al., 2021) and hacktivism (Schneier, 2023) necessitate not only substantial financial investment but also proactive educational interventions to democratise technologies and enhance their accessibility. The activist assemblage in the ouroboros of digital placemaking acts as what Schneier (2023) advocates as the combination of the hacker and avant-garde who follow the rules but break the intent of the system, which differs from the image of a classic Marxist revolutionary who disregards existing rules in attempts to create new ones. The ouroboros therefore describes a system with the propensity, willingness and creativity to tolerate and cultivate its own hacker, which bears a 'politics of negentropy' in Bernard Stiegler's (2018, 2020) term, that creates resistance against the automated and abstracted schema of post-industrialism.

Second, embrace political innovation and policy experimentation. John Dewey's (1981) democratic vision provides an ideal framework for the function of democracy, where citizens and governments collaborate in experimenting ways. Experiments take time and can fail. It is vital to avoid the pitfalls of technological solutionism, which adopts technologies and data to 'fix' urban issues, as it will inevitably trigger what Zuboff (2019, p. 195) calls the 'prediction imperative'. Democracy, inherently experimental, relies on active participation and adaptation to new situations and emerging problems without established formulas or solutions. Like the notion of pharmakon, simultaneously poison and remedy, the digital placemaking ouroboros possesses both generative and destructive potential. The future demands urban publics, experts and governments test new modes of cohabitation and coexistence. These experimental processes and sites will enable us to experience, reflect and learn the essence of becoming conscious, agential power players. Through such engagements, we can reimagine, test and pursue the evolving potentials of digital placemaking.

Finally, re-think Dingpolitik. Graham Harman (2014, pp. 1–4) distinguished the Latourian Dingpolitik from two dominant paradigms: 'Truth Politics' and 'Power Politics'. Truth Politics, characterised by a belief in truth over ignorance, is championed by figures like Alain Badiou and Slavoj Žižek on the left and Plato and Leo Strauss on the right. Conversely, Power Politics, an endless power struggle, is exemplified by postmodern identity politics on the left and the authoritarian tendencies of Hobbesian Leviathan and Carl Schmitt on the right. Both Truth and Power Politics, however, share a dogmatic absolutism and a false dichotomy between human culture and human nature. The range of real-life examples of digital placemaking discussed in this book has revealed the arbitrary rift between the political left and right in a classical sense and demonstrated the possibility of a mode of politics resilient to the intensity and strength of the assembled allies in the actor-networks. To reiterate the principles of Dingpolitik: Politics revolves around issues. Issues mobilise the publics. The mobilised public constituents are in constant flux, marked by interest and relevance rather than purely being defined by geopolitical borders, social class or political parties. A similar mode of quotidian politics, Sachpolitik (issue-related politics), has been conceived and regained popularity in the recent political debate, for example, in Germany, as opposed to Parteipolitik (party politics). Digital placemaking restores the sense of Dingpolitik and constitutes a testing ground for urban democracy. This might allow fostering intermediaries that negotiate processes of inclusion and exclusion and engender new beginnings.

Democracy is an experiment predicated on the notion of participation. There is no established formula or existing solutions to new situations and emerging problems. Like the notion of pharmakon that is both the poison and the remedy, the snake eating its own tail can be as generative as it is destructive. What lies ahead for urban publics, experts and governments alike is testing new ways to inhabit and co-exist. Through these experimental sites and processes, we will experience, reflect on and learn what it means to become aware, agential power players. Similarly, through these occurrences, we can reimagine, test and pursue new potentials of digital placemaking.

Note

1 The Convergence Review (2012), Digital Delivery of Government Services (2018) and Digital Platforms Inquiry (2020–2025).

References

Acemoglu, D., & Robinson, J. A. (2020). *The narrow corridor: States, societies, and the fate of liberty*. New York: Penguin.
Arendt, H. (1958). *The human condition*. Chicago: University of Chicago Press.
Arendt, H. (1961). *The origins of totalitarianism*. London: Penguin Random House.
Blok, A., & Farías, I. (Eds.). (2016). *Urban cosmopolitics: Agencements, assemblies, atmospheres*. London and New York: Routledge.
Brenner, N., Madden, D. J., & Wachsmuth, D. (2011). Assemblage urbanism and the challenges of critical urban theory. *City*, *15*(2), 225–240.
Bruns, A. (2012). Reconciling community and commerce? Collaboration between produsage communities and commercial operators. *Information, Communication & Society*, *15*(6), 815–835.

Chhachhi, A. & Donovan, K. P. (2014). 'Development' as if we have never been modern: Fragments of a Latourian development studies. *Development and Change*, *45*(5), 869–894.
Cresswell, T. (2014). *Place: An introduction*. London: John Wiley & Sons.
Dewey, J., (LW 1-14). 1981–1991. *The Later Works of John Dewey, 1925–1953*, Vol. 14. edited by Jo Ann Boydston. Carbondale: Southern Illinois University.
Dovey, K. (2011). Uprooting critical urbanism. *City*, *15*(3–4), 347–354.
Easterling, K. (2014). *Extrastatecraft: The power of infrastructure space*. London: Verso Books.
Foucault, M. (2003). *"Society Must Be Defended": Lectures at the Collège de France, 1975–1976* (Vol. 1). New York: Macmillan.
Friedman, A. (2023). *The sustainable digital city*. Cham, Switzerland: Springer Nature.
Gabrys, J. (2014). Programming environments: Environmentality and citizen sensing in the smart city. *Environment and Planning D: Society and Space*, *32*(1), 30–48.
Gillespie, T., Boczkowski, P. J., & Foot, K. A. (Eds.). (2014). *Media technologies: Essays on communication, materiality, and society*. Cambridge, MA: MIT Press.
Government of Australia. (1994). Creative Nation: Commonwealth cultural policy, October 1994. https://webarchive.nla.gov.au/awa/20031203235148/, http://www.nla.gov.au/creative.nation/contents.html
Harman, G. (2014). *Bruno Latour: Reassembling the political*. London: Pluto Press.
Heidegger, M (1966). *Discourse on Thinking*. Trans. John M. Anderson and E. Hans Freund. New York: Harper & Row.
Kitchin, R. (2023a). *Digital timescapes: Technology, temporality and society*. Cambridge, UK: Polity Press.
Kitchin, R. (2023b). Urban data power: Capitalism, governance, ethics, and justice. *Data Power in Action: Urban Data in Times of Crisis*, 21, 21–41.
Latour, B. (1990). Technology is society made durable. *The Sociological Review*, *38*(S1), 103–131.
Latour, B. (2004). Why has critique run out of steam? From matters of fact to matters of concern. *Critical Inquiry*, *30*(2), 225–248.
Levenda, A. M., Keough, N., Rock, M., & Miller, B. (2020). Rethinking public participation in the smart city. *The Canadian geographer/Le géographe canadien*, *64*(3), 344–358.
Locke, J. (1689/1967). *Two treatises of government*. Cambridge, MA: Cambridge University Press.
Marcuse, H. (1964) *One-dimensional man: Studies in the ideology of advanced industrial society*. Boston: Beacon Press.
Massey, D. (2005) *For space*. London: Sage.
McLuhan, M. (1967). The invisible environment: The future of an erosion. *Perspecta*, 11, 163–167.
Mill, J. S. (1977). "Considerations on Representative Government", *Essays on Politics and Society Collected Works of John Stuart Mill*, vol. XIX. Toronto: University of Toronto Press/Routledge & Kegan Paul, pp. 371–577.
Mumford, L. (1937/2011). "What is a City?": Architectural Record (1937). In: R. LeGates & F. Stout (eds). *The City Reader* (pp. 123–127). London and New York: Routledge.
Murphy, Michelle. 2015. Unsettling care: Troubling transnational itineraries of care in feminist health practices. *Social Studies of Science*, *45*, 717–737.
O'Brien, D. T. (2018). *The urban commons: How data and technology can rebuild our communities*. Cambridge, MA: Harvard University Press.
Paddison, R., Philo, C., Routledge, P., & Sharp, J. (2002). *Entanglements of power: Geographies of domination/resistance*. New York and London: Routledge.
Rose, G. (2022). *Seeing the city digitally: Processing urban space and time*. Amsterdam: Amsterdam University Press.

Saitta, D. (2024). *First cities: Planning lessons for the 21st century*. Cambridge, UK: Cambridge University Press.
Schneier, B. (2023). *A hacker's mind: How the powerful bend society's rules, and how to bend them back*. New York: W. W. Norton & Company.
Schwennesen, N. (2019). Algorithmic assemblages of care: Imaginaries, epistemologies and repair work. *Sociology of Health & Illness*, *41*, 176–192.
Sharpe, L. J. (1970). Theories and values of local government. *Political Studies*, *18*(2), 153–174.
Simone, A. (2010). *City life from Jakarta to Dakar: Movements at the crossroads*. London and New York: Routledge.
Stiegler, B. (2018). *The neganthropocene*. London: Open Humanities Press.
Stiegler, B. (2020). *Nanjing Lectures (2016–2019)*. London: Open Humanities Press.
Townsend, A.M. (2013). *Smart cities*. New York: W. W. Norton & Company.
Virilio, P. (1986). *Speed and politics: An essay on dromology* (M. Polizzotti, Trans.). New York: Columbia University Press.
Vohland, K., Land-Zandstra, A., Ceccaroni, L., Lemmens, R., Perelló, J., Ponti, M., … Wagenknecht, K. (2021). *The science of citizen science*. Cham: Springer Nature.
Weyl, E. G., & Tang, A. (2024). *Plurality: The future of collaborative technology and democracy*. RadicalXChange.
Zhao, F., Fashola, O. I., Olarewaju, T. I., & Onwumere, I. (2021). Smart city research: A holistic and state-of-the-art literature review. *Cities*, *119*, 103406.
Zuboff, S. (2019). *The age of surveillance capitalism: The fight for a human future at the new frontier of power*. London: Profile Books.

Index

Actor-Network Theory 17; ANT-informed ethnography 17–18
aesthetic 31, 50, 54–56, 65, 86, 104, 107, 122
Agamben, G. 82
Alibaba 3
Alphabet 1, 6, 106–107
Apple: store 8, 19; Square 9, 17, 94–96
Arendt, H. 17, 25, 40, 59, 84, 124–126
Arnstein, S. R. 10, 28, 111; Arnstein's Ladder 10–11, 43, 85, 102–104
assemblage: activist 12, 15, 17, 40–41, 65–67; organisational 13, 17, 40–42, 71–72, 85–88; governmental 9, 13, 15, 17, 39–41, 65, 94, 111; urban 8–9, 37–38

Barns, S. 4, 32, 106
Bauman, Z. 9
Bourriaud, N 31
Bratton, B 2, 8
Brenner, N. 10, 21, 38, 60, 120

Cardullo, P. 10–11, 43, 85, 100, 102
Carmona, M. 30, 42
Club of Rome 4
Cresswell, T. 6, 121

Deleuze, G. 6, 37
democracy 5, 7–8, 10, 14, 29–30; digital democracy 34–37, 108, 125; urban democracy 43, 105, 128
digital city, digital cities 17, 50, 72
Dingpolitik 39, 127–128
Dovey, K. 37–38, 119

emergence 3, 5, 12–13, 26, 39, 43, 121–122
engagement portal 17, 33, 92–93, 97, 99, 121

Facebook 52, 55
Fung, A 43, 103

G0v 20, 34, 94, 108
Gehl, J 26–27
Google 32, 55–56, 64, 106
Groys, B. 56

Habermas, J. 2, 25, 39, 103
design hackathon 17, 72, 83, 87, 121–123; Taiwanese Presidential Hackathon 108–109
Halegoua, G. R. 31, 35–36
Harman, G. 14, 119, 127
Harvey, D. 66, 70, 83
Heidegger, M. 122

International Association of Public Participation (IAP2) 21, 28, 93, 103, 105
Isabelle, S. 12

Jacobs, J. 26, 28, 59

Kitchin, R. 32, 34–35, 43, 85, 100, 102

Latour, B. 13–14, 18–20, 38–41, 55, 119, 122, 125, 127
Lefebvre, H. 4, 10, 26, 29–30, 58, 107

Mannheim, K. 41
Mattern, S. 6, 71, 83
McLuhan, M. 122
McQuire, S. 5–6, 36, 81, 107
McShane, I. 93
Mumford, L. 2, 122

negentropy 66, 127

oligopticon 20, 40
ouroboros of digital placemaking 12–17, 38, 116–120, 125, 127

Papastergiadis, N. 14, 104
participatory condition 7–9, 42
pharmakon 8, 11, 127–128
place: branding 11, 121; placemaking 3, 25–29; digital placemaking 4–5, 30–39; creative placemaking 3, 27, 70
politics: political circle 13, 38–39, 43; power 127–128; truth 4, 127; of digital placemaking 2, 4, 39, 43

Sloterdijk, P. 4
smart city 3–6, 32–34, 43, 118
socio-technical natality 17, 19, 124–126
sovereignty 2, 8–9, 107
Stiegler, B. 8, 66, 127
Sunflower Movement 34, 108, 110

Tang, A. 37, 108
techno-capitalism 63, 72, 107
temporality 122
Thrift, N. 40, 72
Townsend, A. 75, 127
Tuan, Y. 6

urban cosmopolitics 8, 12

Virilio, P. 118

Whyte, W. H. 26–27
Wilken, R. 31
Winner, L. 93

Zuboff, S. 55, 71, 126
Zumthor, F. 54

FSC
www.fsc.org
MIX
Papier | Fördert gute Waldnutzung
FSC® C083411

Kolibri 360 GmbH
Ferdinand-Jühlke-Straße 7
99095 Erfurt, Deutschland
produktsicherheit@kolibri360.de